土木工程学术研究丛书

U0725578

Adaptive Finite Element Method of Eigenproblems in Beam and Shell

梁壳特征值问题自适应有限元法

王永亮　著

Yongliang Wang

中国建筑工业出版社

CHINA ARCHITECTURE & BUILDING PRESS

图书在版编目（CIP）数据

梁壳特征值问题自适应有限元法 = Adaptive Finite Element Method of Eigenproblems in Beam and Shell：英文 / 王永亮著. — 北京：中国建筑工业出版社，2023.3

（土木工程学术研究丛书）

ISBN 978-7-112-28352-1

Ⅰ. ①梁… Ⅱ. ①王… Ⅲ. ①特征值问题-有限元法-英文 Ⅳ. ①O175.9

中国国家版本馆 CIP 数据核字（2023）第 019310 号

责任编辑：刘瑞霞 梁瀛元
责任校对：张辰双

土木工程学术研究丛书
Adaptive Finite Element Method of Eigenproblems in Beam and Shell
梁壳特征值问题自适应有限元法
王永亮 著
Yongliang Wang

*

中国建筑工业出版社出版、发行（北京海淀三里河路 9 号）
各地新华书店、建筑书店经销
北京鸿文瀚海文化传媒有限公司制版
北京建筑工业印刷厂印刷

*

开本：787 毫米×1092 毫米 1/16 印张：14 字数：345 千字
2023 年 1 月第一版 2023 年 1 月第一次印刷
定价：**78.00** 元
ISBN 978-7-112-28352-1
（40784）

Adaptive Finite Element Method of Eigenproblems in Beam and Shell

To My Teachers and Students

Abstract

The eigenproblems investigated here include significant problems in structural engineering fields such as free vibration, elastic stability, buckling, and damage-induced disturbance in beam and shell structures, as well as mathematically challenging vector Sturm-Liouville eigenproblems. The eigenproblems belong to a class of typical nonlinear problems, which is challenging to obtain high-precision continuous order eigenvalues and eigenfunctions with high efficiency. The accuracy and efficiency of solutions put forward high requirements for numerical methods. This book develops high-performance h-and hp-version adaptive finite element algorithms and methods to obtain high-precision solutions involved in vibration, stability, and damage disturbance of beam and shell. The book covers the following main contents: (1) adaptive finite element method for vector Sturm-Liouville eigenproblems, (2) adaptive finite element method for vibration of non-uniform and variable curvature beams, (3) adaptive finite element method for vibration disturbance of cracked beams, (4) adaptive finite element method for damage detection of cracked beams, (5) adaptive finite element method for stability disturbance of cracked beams, (6) adaptive finite element method for vibration of cylindrical shells, (7) improved hp-version adaptive finite element method for vibration of cylindrical shells, and (8) adaptive finite element method for vibration disturbance of cracked cylindrical shells.

Given its scope, the book offers a valuable reference guide for researchers, postgraduates and undergraduates majoring in engineering mechanics, computational mathematics, and civil engineering.

Introduction of the author

Dr. Yongliang Wang is currently a researcher in the Department of Engineering Mechanics, School of Mechanics and Civil Engineering, State Key Laboratory of Coal Resources and Safe Mining, at China University of Mining and Technology (Beijing), and the head of computational mechanics group. He obtained his Ph. D. degree from the Department of Civil Engineering at Tsinghua University in 2014. In 2015, 2016, 2017, and 2019, he successively visited the Zienkiewicz Centre for Computational Engineering at Swansea University, UK, the Applied and Computational Mechanics Center at Cardiff University, UK, and the Rockfield Software Ltd, UK, to carry out cooperative research. In 2022 and 2023, he visited University of California, Berkeley and San Diego, USA, as visiting scholar.

His research interests include high-performance adaptive finite element method, computation and analysis of rock damage and fracture, and structural vibration and stability. He has also taught computational mechanics at the undergraduate level and the basic theory of the finite element method, computational solid mechanics, and rock fracture mechanics, and frontier and progress in mechanics at the graduate level. He is a board member of the Soft Rock Branch of the Chinese Society for Rock Mechanics and Engineering and a member of the Chinese Society of Theoretical and Applied Mechanics, China Civil Engineering Society, and China Coal Society; moreover, he serves as a project expert of the National Natural Science Foundation of China, and project expert of the DegreeCenter of Ministry of Education of China.

As the person in charge, he presided over eighteen research projects, including the National Natural Science Foundation of China, Beijing Natural Science Foundation, Teaching Reform and Research Projects of Undergraduate Education, the China Postdoctoral Science Foundation, Fundamental Research Funds for the Central Universities, Ministry of Education of China, Key Laboratory Open Project Foundation of Soft Soil Characteristics and Engineering Environment, and Yue Qi Young Scholar Project Foundation, CUMTB. He participated in twelve research projects, including the Major Scientific Research Instrument Development of the National Natural Science Foundation and the National Key Research and Development Program of China. He published 4 books in English and more than 100 academic papers; furthermore, he obtained more than 50 software copyrights. He has received the Rock Mechanics Education Award, Chinese Society for Rock Mechanics and Engineering (2021), the Excellent Supervisor Award, CUMTB (2019, 2020), the Science and Technology Award, China Coal Industry

Association (2019), the Yue Qi Young Scholar Award, CUMTB (2019), the Emerald Literate Highly Commended Paper Award (2018), and the Frontrunner 5000 (F5000) Top Articles in Outstanding S&T Journals of China (2016, 2023).

Preface

Dynamic failure and instability are the main failure modes of structures. Compared with static failure, dynamic failure has shorter failure duration and greater destructive capability. The frequency in free vibration analysis of structure is the basis to avoid resonance failure and evaluate forced vibration behaviour, besides, instability buckling analysis is the premise to avoid stability failure. The damage in the structures will disturb these vibrations and instabilities, resulting in more complex dynamic behaviors. The free vibration and elastic stability buckling problems in engineering here belong to a class of nonlinear problems in mathematics, namely eigenproblems. The eigenvalue problems include continuous order eigenvalues and eigenfunctions. Natural frequency and buckling load belong to the category of eigenvalues, and vibration mode and buckling mode belong to the category of eigenfunctions. High performance numerical methods and algorithms are required to efficiently obtain high-precision eigenvalue solutions in engineering field.

This book is organized as follows. Chapter 1introduces the research background and significance. In Chapter 2, the adaptive finite element method for vector Sturm-Liouville eigenproblems, is introduced. In Chapter 3, the adaptive finite element method for vibration of non-uniform and variable curvature beams, is introduced. In Chapter 4, the adaptive finite element method for vibration disturbance of cracked beams, is introduced. In Chapter 5, the adaptive finite element method for damage detection of cracked beams, is introduced. In Chapter 6, the adaptive finite element method for stability disturbance of cracked beams, is introduced. In Chapter 7, the adaptive finite element method for vibration of cylindrical shells, is introduced. In Chapter 8, the improved hp-version adaptive finite element method for vibration of cylindrical shells, is introduced. In Chapter 9, the adaptive finite element method for vibration disturbance of cracked cylindrical shells, is introduced. Finally, Chapter 10 summarizes the main conclusions and prospects of this book.

The author gratefully acknowledges the financial supports fromthe research projects led by the author, i. e. the National Natural Science Foundation of China (Grant Nos. 41877275 and 51608301), the Beijing Natural Science Foundation (Grant L212016), the China Postdoctoral Science Foundation (Grant Nos. 2018T110158, 2016M601170, and 2015M571030), the Key Laboratory Open Project Foundation of Soft Soil Characteristics and Engineering Environment (Grant No. 2017SCEEKL003), the Fundamental Research Funds for the Central Universities, Ministry of Education of China (Grant No. 2019QL02), the Teaching Reform and Research Projects of Undergraduate Education, CUMTB (Grant Nos. J210613,

J200709, and J190701), the Innovation Training Projects for Undergraduates, CUMTB (Grant Nos. 202106001, 202106030, C202006976, and C201906327), and the Yue Qi Young Scholar Project Foundation, CUMTB (Grant No. 190618).

The author gratefully acknowledges the guidance and advice from the respectable tutors during the master, Ph. D. , and postdoctorate stages, Prof. Yuan Si and Prof. Zhuang Zhuo of Tsinghua University, and Prof. Wu Jianxun and Prof. Ju Yang of the China University of Mining and Technology (Beijing). During the postdoctorate stage, the author visited several research centres for computational mechanics in famous foreign universities as visiting scholar; the author gratefully acknowledges the advice and comments from the collaborators, Prof. Li Chenfeng, Prof. Feng Yuntian, and Prof. D. Roger J. Owen of the Zienkiewicz Centre for Computational Engineering at Swansea University in the UK, Prof. David Kennedy and Prof. Frederic W. Williams of the Applied and Computational Mechanics Group at Cardiff University in the UK, John Cain, Melanie Armstrong, and Fen Paw at Rockfield Software Ltd. in the UK, and Prof. Robert L. Taylor of the Department of Civil and Environmental Engineering at the University of California, Berkeley, in the USA. The author gratefully acknowledges the participation and work from the Ph. D students, including Nana Liu, Xuguang Liu, Yishuo Cui, Ruiguang Feng of the China University of Mining and Technology (Beijing). The author also gratefully acknowledges the editors, Ms Yingyuan Liang and Ms Ruixia Liu, at the China Architecture & Building Press in China, for providing many suggestions and much assistance on formatting modifications and typesetting adjustments for improving this manuscript.

The key contents of this book, such as vibration, stability, and damage disturbance, are crucial in dynamic behaviors in beam and shell and are challenging the researchers' best knowledge. Further work on these fields is needed for both theoretical and algorithm advancements. Because this book is restricted by the limited knowledge of the author, a few errors are unavoidable. The author hopes all that experts, scholars, and other readers of this book will provide helpful suggestions for the book's improvement.

Dr. Yongliang Wang
Department of Engineering Mechanics
School of Mechanics and Civil Engineering
State Key Laboratory of Coal Resources and Safe Mining
China University of Mining and Technology (Beijing)
D11 Xueyuan Road, Beijing, 100083, China
Homepage: www. wangyongliang. net
Email: wangyl@cumtb. edu. cn
January, 2022

Catalogue

Chapter 1
Introduction

1. 1　Introduction

The eigenproblems investigated here include significant problems in structural engineering fields such as free vibration, elastic stability, buckling, and damage-induced disturbance in beam and shell structures, as well as mathematically challenging vector Sturm-Liouville eigenproblems. The eigenproblems belong to a class of typical nonlinear problems, which is challenging to obtain high-precision continuous order eigenvalues and eigenfunctions with high efficiency. The accuracy and efficiency of solutions put forward high requirements for numerical methods. The accuracy and effectiveness of finite element solutions depend on mesh quality. A few high-performance adaptive algorithms have been proposed to automatically optimise the mesh based on the complexity of the problem. Compared with the traditional method, the adaptive finite element algorithms exhibit improved accuracy and efficiency[1, 2]. It should be noted that adaptive algorithms have some obvious advantages over traditional numerical methods in terms of solving challenging problems, such as fracture, singularity, and eigenvalue problems. The following will focus on the basic procedures and categories of high-performance adaptive finite element method, as well as the increasing number of relevant studies in recent years, to show the potential of this new methods. This chapter will not introduce too much reference information, but will summarize the relevant research progress in the following chapters of each subtopic.

1. 2　Basic procedures and categories of high-performance adaptive finite element method

The high-performance adaptive finite element method can be implemented by the following three-step adaptive procedures[3], as shown in Figure 1. 1:

（1）**FE (finite element) solutions.** On the current mesh, the finite element solution u^h is obtained by conventional finite element method.

（2）**Errorestimation.** Based on the finite element solution, the superconvergent solution u^* is computed through superconvergent patch recovery displacement. Subsequently, u^* is used to estimate the error of the u^h, and then the errors in energy norm on the global domain are marked out.

（3）**Solution refinement.** If the pre-specified error toleranceis not satisfied for the elements above, the mesh is further subdivided to generate a new mesh based on the computed errors, and the mesh refinement procedure is used to refine the current mesh on the global domain. Then, the procedure returns to the first step (i. e. , FE solutions) and the cycle repeats until all the elements satisfy the pre-specified error tolerance.

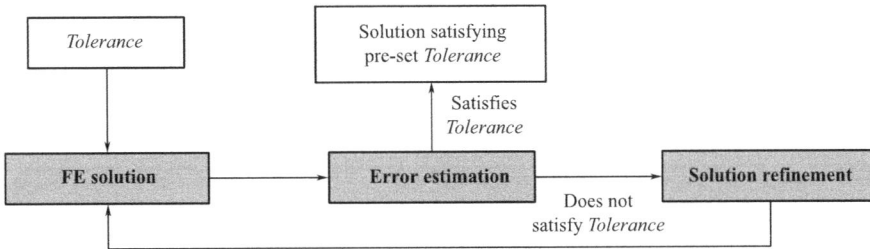

Figure 1.1 Basic process flow of the high-performance adaptive finite element algorithms

There are various procedures for the refinement of finite element solutions, and they broadly fall into three categories:

(1) **h-version refinement:** The same order of element polynomial isused, but the elements are changed in size (in some locations they are made larger, and in others smaller) to achieve maximum efficiency in reaching the desired solution.

(2) **p-version refinement:** The procedure uses the same element size and simply increases the order of the element polynomials.

(3) **hp-version refinement:** The procedure combinesthe h-and p-version refinements. Inthisprocedure, both the element size and the degree of the element polynomial are altered.

In this book, the h-, and hp-version adaptive finite element method are developed for eigenproblems in beam and shell, containing the vibration, stability, and damage disturbance.

1.3 Applications of adaptive finite element method in eigenproblems of beam and shell: Vibration, stability, and damage disturbance

Figure 1.2 shows the number of academic publications of adaptive finite element

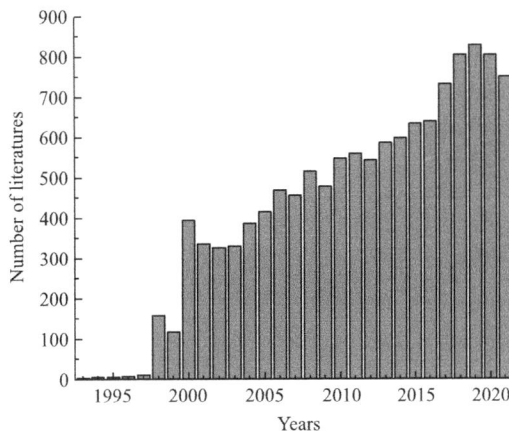

Figure 1.2 Number of literatures based on adaptive finite element method

method in database of Web of Science. It can be seen that since the 1990s, the number of papers published by adaptive finite element method has increased sharply. In recent years, nearly 1000 papers have been published every year. It shows that the research of adaptive finite element method is concerned and used.

Figures 1.3-1.5 show the academic papers published by the adaptive finite element method in the research field of structural eigenvalues such as free vibration, elastic stability and damage identification this year. It can be seen that since the 1990s, the number of papers published by the adaptive finite element method has increased significantly. In recent years, nearly 60, 100, and 40 papers on vibration, stability, and damage disturbance have been published every year. It shows that the adaptive finite element method has attracted much attention and use in these fields.

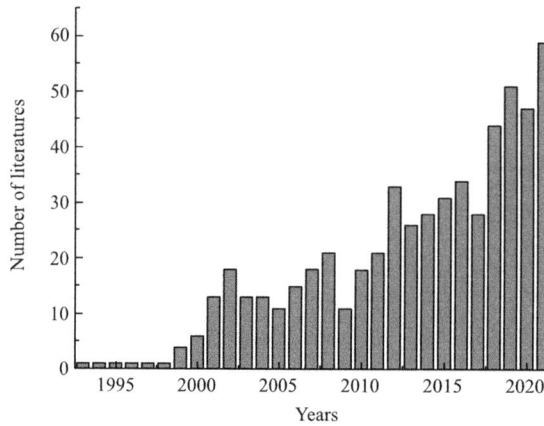

Figure 1.3   Number of literatures in free vibration of
structures based on adaptive finite element method

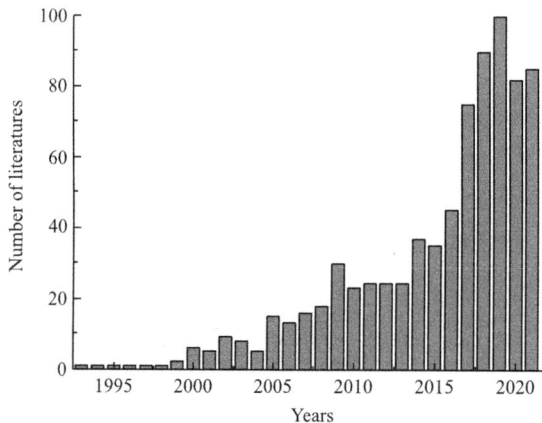

Figure 1.4   Number of literatures in elastic stability of
structures based on adaptive finite element method

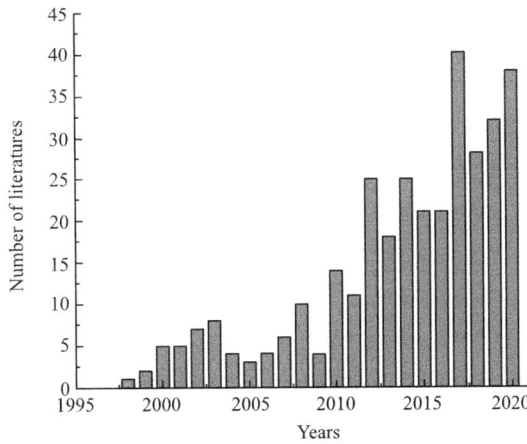

Figure 1. 5   Number of literatures in damage disturbance of
structures based on adaptive finite element method

1. 4   Research aims and contents of the book

This book develops high-performance h-and hp-version adaptive finite element algorithms and methods to obtain high-precision solutions involved in vibration, stability, and damage disturbance of beam and shell. The book covers the following main research aims and contents:

(1) Adaptive finite element method for vector Sturm-Liouville eigenproblems.

(2) Adaptive finite element method for vibration of non-uniform and variable curvature beams.

(3) Adaptive finite element method for vibration disturbance of cracked beams.

(4) Adaptive finite element method for damage detection of cracked beams.

(5) Adaptive finite element method for stability disturbance of cracked beams.

(6) Adaptive finite element method for vibration of cylindrical shells.

(7) Improved hp-version adaptive finite element method for vibration of cylindrical shells.

(8) Adaptive finite element method for vibration disturbance of cracked cylindrical shells.

This book mainly involves the following characteristics, e. g. basic English contents, frontier researches, representative and typical examples. The specific contents are as follows:

(1) **Basic English contents:** the contents of the book are mainly in English, using simple and common English words to assist the interpretation of key professional words, reduce the requirements of readers in China (or non-native English speaking countries) for all English words, and meet the needs of domestic scholars, graduate students and

undergraduate students for academic research and learning in professional English.

(2) **Frontier academic researches:** the book introduces the error estimation and adaptive method of finite element solution, and focuses on the innovative micro-crack model technology to characterize the multi-crack damage conditions, Sturm sequence J-counting method to ensure that the continuous order frequency does not lose eigenpair, and the inverse power (subspace) iterative solution method to solve single eigenpair and multiple eigenpairs, so as to make readers understand the frontier academic researches of 13 computational mechanics such as high-performance computation and analysis technology of adaptive finite element method.

(3) **Verification and validation using representative and typical examples:** the book systematically introduces the eigenproblems of multiple representative and typical of beams and shells, including vector Sturm-Liouville eigenproblems, vibration of non-uniform variable curvature beam, vibration disturbance of beams with crack damage, damage detection of beams with crack damage, stability disturbance of beams with crack damage, vibration of cylindrical shells, vibration disturbance of cylindrical shells with crack damage. A variety of typical numerical examples are provided for each type of problem to verify the effectiveness of the proposed adaptive finite element method.

References

[1] Zienkiewicz O C. The background of error estimation and adaptivity in finite element computations [J]. Computer Methods in Applied Mechanics & Engineering, 2006, 195 (4-6): 207-213.

[2] Zienkiewicz O C, Taylor R L, Zhu J Z. The Finite Element Method: Its Basis and Fundamentals [M]. 7th edition. Elsevier Press, 2013.

[3] Wang, Y L. Adaptive analysis of damage and fracture in rock with multiphysical fields coupling [M]. Springer Press, 2021.

Chapter 2
Adaptive finite element method for vector Sturm-Liouville eigenproblems

2. 1　Introduction

General eigenproblems in system of ordinary differential equations (ODEs) serve as mathematical models for vector Sturm-Liouville (SL) and free vibration problems of structural members[1, 2]. Mathematically, the vector SL problem is a special case of eigenproblem[3] regarded as the basic theoretical equation and modelgoverning physical processes and mechanical behaviour such as the competition and exclusion process of biological populations, river bed migration, or droplet diffusion on solid surfaces[4, 5]. The free vibration problems of structural members may be related to the following categories, i. e. free vibration of straight Timoshenko beams, planar curved beams and moderately thick circular cylindrical shells[6, 7]. The development of general analytical methods for eigenproblems in system of ODEs can provide solutions for vector SL and free vibration problems. Further, high-precision solutions of eigenvalues and eigenfunctions play important roles in research and engineering. For example, the representative dynamic analysis of structures based on high-precision solutions of natural frequency (eigenvalue) or vibration mode (eigenfunction)[8, 9] enables the accurate identification of the number, size, and location of the damages in cracked structures[10, 11]. However, solutions that meet user-specified error tolerances are challenging to provide, especially for issues involving the coefficients of variable matrices[12], coincident and adjacent approximate eigenvalues[13], continuous orders of eigenpairs[14], varying boundary conditions[15], and variable cross-sections or curvatures[16-18]. To overcome problems such as those listed above, a general analysis method that provides high-precision solutions has to be developed.

The theoretical solutions forgeneral eigenproblems in system of ODEs are difficult to obtain. Some theoretical methods, such as the dynamic stiffness method[17], are applicable only to simple geometrics and boundary conditions, through model simplification. Some numerical methods and models are used as alternative technical means. Toimprove the accuracy of the finite element (FE) solutions for second order SL problems, a scheme applyingspecial FEmodels[19] and asymptotic correction techniques[20] have been proposed. For regular and singular vector SL problems, spectral function method[21], shooting method[22], and functional-discrete method[23] were developed. In order to acquirethe high-precision solutions of eigenvalues, special procedures based on Hamiltonian forms, such asSL11F, SL12F[24] and components of the NAG library[25], have also been developed. To compute solutions for frequencies and modes of the structures, effective analysis strategies, such as inverse and subspace iterations in the conventional finite element method (FEM)[26] and semi-analytic ODE solver[27], have been developed. However, these methods and procedures do not impose error control on both of eigenvalues and eigenfunctions and, hence, none of them serves as complete solvers.

Recently, error estimators and adaptive analysis techniques have been proposed for eigenproblems. To estimate errors of eigenpairs solutions, and a *posteriori* error estimator of a multirate numerical method for ODEs[28] and a goal-oriented *posteriori* error estimator of multiscale PDE-ODE coupled systems[29] have been proposed. Some adaptive analysis strategies were also proposed, for example, anadaptive solution of boundary value problems (BVPs) in singularly perturbed second order ODEs by combining the extended Numerov method with an iterative local h-refinement[30], and an adaptive ODE solver using the extended Kalman filtering algorithms[31]. The solution accuracy of the conventional FEM depends on an appropriate mesh selection[32]. The combination of conventional FEM and adaptive mesh refinement techniques demonstrates a good computation effectiveness and an improved accuracy of the solutions. Through methods such as local modelling error estimation and goal-oriented adaptivity[33, 34] or *posteriori* error estimates and induced adaptive methodsbased on superconvergent patch recovery[35, 36], complex heterogeneous materials related BVPs have been solved more effectivelyandreliablythan by conventional FEM. However, these procedures for adaptive FEM are not aimed atgeneral eigenproblems in systems of ODEs and almost impossible to find corresponding reliable eigenpairs to satisfy the pre-specified error tolerance.

To solve general eigenproblems in systems of second order ODEs, this study proposes an h-version adaptiveFEM based on the superconvergent patch recovery displacement method. Uniform refined meshes were established to suit the changes in the eigenfunctions, and the computed results satisfied the pre-specified error tolerance. The chapter is organised as follows. We present the basic adaptive approach and the procedure for eigenproblems in systems of ODEs in Section 2.2. FE eigenpairs solutions by inverse and subspace iteration techniques are introduced in Section 2.3. In Section 2.4 and 2.5, key techniques used in adaptive analysis procedure, such as error estimation for eigenfunctions and h-version mesh refinement, are presented. Representative numerical examples are presented in Section 2.6, to show the performance of the proposed method and algorithm. Finally, the main conclusions are summarised in Section 7.

2.2 Adaptive approach for eigenproblems in systems of ordinary differential equations

2.2.1 Eigenproblems in system of second order ordinary differential equations

The eigenproblem solved in this study consists of findingthe eigenvalues λ and the associated n_d dimensional vector eigenfunctions $\boldsymbol{u}(x) = (u_1(x), \cdots, u_{n_d}(x))^{\mathrm{T}}$ for the following systems of second order ODEs (Greenberg, 1991; Kurochkin, 2014):

$$\boldsymbol{L}_2 \boldsymbol{u} \equiv -(\boldsymbol{A}\boldsymbol{u}' + \boldsymbol{B}\boldsymbol{u})' + \boldsymbol{B}^{\mathrm{T}}\boldsymbol{u}' + \boldsymbol{C}\boldsymbol{u} = \lambda \boldsymbol{R}\boldsymbol{u}, \ a < x < b \tag{2.1}$$

where the prime mark ($'$) denotes the derivative with respect to the independent variable x, L_2 is the associated self-adjoint second order differential operator, and A, B, C, and R are continuous $n_d \times n_d$ matrix functions on (a, b). A, C and R are symmetric, A and R are positive definite, and B^T is transposed matrix of B.

Mathematically, the vector SL problem is a special case of the ODEs eigenproblem defined in Equation (2.1), with $B = 0$. Structural vibration problems are chosen to illustrate the possible physical interpretations of the equations. The structural natural frequencies ($\omega = \sqrt{\lambda}$) and modes from the structural vibration problems are corresponding to the eigenvalues and vector eigenfunctions, respectively.

The boundary conditions of the eigenproblems for Equation (2.1) are:

$$\boldsymbol{u}(a) = \mathbf{0}, \ \boldsymbol{u}(b) = \mathbf{0} \tag{2.2}$$

In this study, these fixed boundary conditions are taken as the default values, without losing generality. However, boundary conditions such as the hinged (simply supported) boundary condition can be easily introduced, when necessary.

2.2.2 Stop criterion and adaptive procedure

The computation aim of the proposed method is to find the eigenpair solutions of continuous orders, $(\lambda_k^h, \boldsymbol{u}_k^h)(k = 1, \cdots, n)$, on sufficiently fine meshes, π_k ($k = 1, \cdots, n$), such that:

$$|\lambda_k - \lambda_k^h| \leqslant Tol \cdot (1 + |\lambda_k|) \tag{2.3}$$

$$\max_{a < x < b} |u_{i,k}(x) - u_{i,k}^h(x)| \leqslant Tol \text{ with } \max_i (\max_{a < x < b} |u_{i,k}(x)|) = 1, \ i = 1, \cdots, n_d \tag{2.4}$$

where $(\lambda_k, \boldsymbol{u}_k)(k = 1, \cdots, n)$ are the exact solutions and Tol is the pre-specified error tolerance for both eigenvalues and eigenfunctions.

Although the above criteria are the ultimate aims, they cannot be used as the stop criteria for a computing procedure, because exact solutions are not usually available. On the current mesh, the superconvergent solution $\boldsymbol{u}_k^*$ which possesses higher accuracy than conventional FE solution, can be used instead of the exact solution to estimate the accuracy of the eigenfunction in energy norm[9-38]. In this case, the superconvergent solution of the eigenvalue is obtained through computing the Rayleigh quotient[39, 40] by using the superconvergent solutions of the eigenfunction. It should be noted that the eigenvalue solution has a higher convergence order and accuracy than the eigenfunction solution, and, therefore, the stop criterion can be formed by error estimation of the eigenfunction. Thus, an eigenfunction solution that satisfies the pre-specified error tolerance and further guarantees the high-precision accuracy of the eigenvalue can be obtained.

The flow chart of the adaptive FEM procedure for eigenproblems in system of second order ODEs can be summarised, as shown in Figure 2.1. First, the input parameters, coefficients, boundary conditions, n orders, initial mesh and pre-specified error tolerance

Tol are defined. Subsequently, the solution computation procedure starts from the lowest order eigenpair (λ_1, $\boldsymbol{u}_1$), and then successively advances until the highest order eigenpair (λ_n, $\boldsymbol{u}_n$) has been found. To start with the first eigenpair, an initial mesh π_0 must be specified by the user. After the adaptive solution for the first eigenpair has been completed, the resulting final mesh π_1 is used as the initial mesh for computing the next order eigenpair. This procedure is repeated until the highest order eigenpair has been computed.

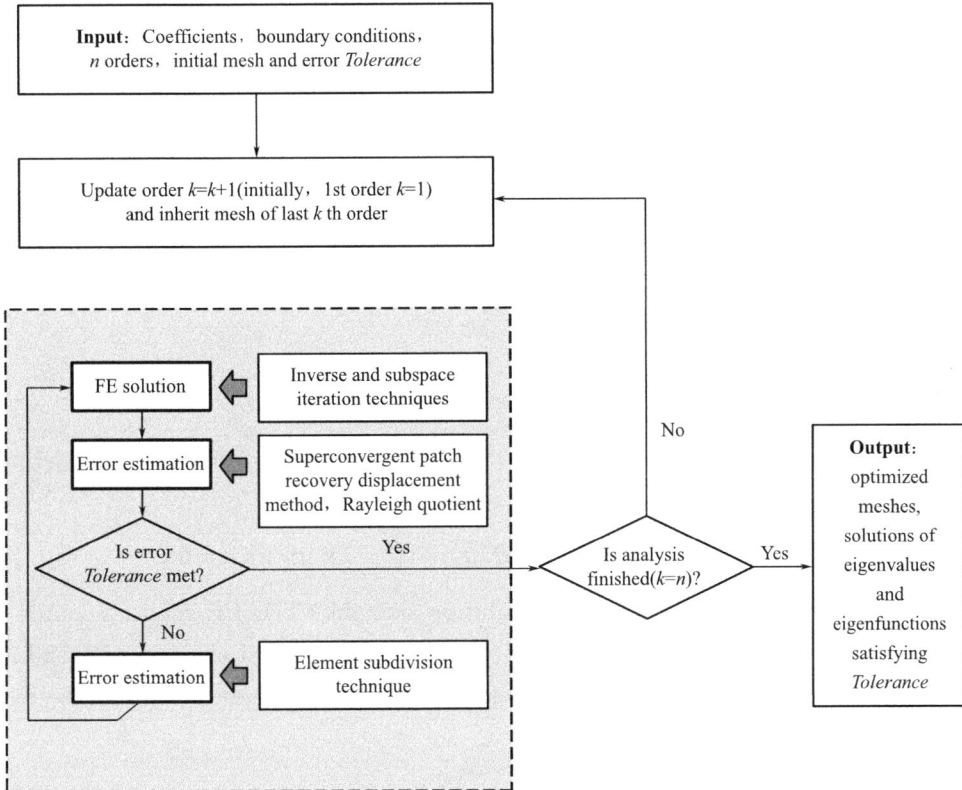

Figure 2. 1   Flow chart of the adaptive FEM procedure
for eigenproblems in system of second order ODEs

The proposed adaptive procedure achievesthe desired results (pre-specified error tolerance, high-precision accuracy) for each eigenpair (λ, $\boldsymbol{u}$), by simply implementing the following three-step adaptive procedure:

（1）**FE eigenpairs solutions.** On the current mesh, the FE solution (λ^h, $\boldsymbol{u}^h$) is obtained by conventional FEM, using the inverse iteration for single eigenpairs and the subspace iteration for coincident eigenpairs.

（2）**Error estimation for eigenfunctions.** Based on the FE solution (λ^h, $\boldsymbol{u}^h$) found, the superconvergent solution $\boldsymbol{u}^*$ of the eigenfunction is computed through superconvergent patch recovery displacement, and the superconvergent solution of the eigenvalue λ^*, through the Rayleigh quotient. Subsequently, $\boldsymbol{u}^*$ is used to estimate the error of the $\boldsymbol{u}^h$, and then the errors in energy norm on the global domain are marked out.

(3) **h-version mesh refinement.** If *Tol* is not satisfied for the elements above, the mesh is further subdivided to generate a new mesh based on the computed errors, and the mesh refinement procedure is used to refine the current mesh on the global domain. Then, the procedure returns to the first step (i. e. , the FE eigenpairs solutions) and the cycle repeats until all the elements satisfy the pre-specified error tolerance.

The steps listed above constitute a round of adaptive iteration. After a series of such iterations, the procedure may further refine and optimise the mesh and solutions satisfying the pre-specified error tolerance. On the new refined mesh, a round of inverse or subspace iterations is implemented to form adaptive inverse or adaptive subspace iterations, respectively. Finally, the refined meshes for all orders of eigenpairs are outputted, and the solutions of the eigenvalues and eigenfunctions on the optimised meshes satisfy the pre-specified error tolerance.

2.3　Finite element solutions of eigenpairs

2.3.1　Finite element discretisation

The weak form for the eigenproblems in system of second order ODEs as defined in Equation (2. 1) can be expressed as:

$$\int_{\Omega} \{v'^{\mathrm{T}}(Au' + Bu) + v^{\mathrm{T}}[B^{\mathrm{T}}u' + (C - \lambda R)u]\} \, \mathrm{d}x = 0 \tag{2.5}$$

where v is trial function, and Ω is the solution domain. The FE model adopted uses conventional degree m polynomial elements. Let e denote a typical element with end nodes coordinates $\overline{x}_1$, $\overline{x}_2$ and length h. Let the trial function on an element of degree m be written as[26]:

$$v = \sum_{i=1}^{m+1} N_i v_i \tag{2.6}$$

Here, N_i is a $n_d \times n_d$ shape function matrix defined by:

$$N_i = N_i I, \ i = 1, 2, \cdots, m+1 \tag{2.7}$$

where I is the $n_d \times n_d$ identity matrix.

Utilising the conventional FEM, the element stiffness and mass matrices (K^e and M^e, respectively) are computed and assembled to form the global stiffness and mass matrices K and M. The FE equation can be derived as an eigenvalue equation in the following matrix form:

$$KD = \lambda MD \tag{2.8}$$

where D is the eigenfunction vector, and the matrices K and M are independent of λ. Given an arbitrary trial value λ_a as the shift value, Equation (2. 8) can be equivalently written in the shifted form[26]:

$$K_a D = \mu MD \text{ with } K_a = K - \lambda_a M, \ \mu = \lambda - \lambda_a \tag{2.9}$$

In the proposed method, the convectional FE computation foreigenpair solutions is based on the Sturm sequence property[39, 40], which can be expressed as:

$$K - \omega^2 M = LD(\omega)L^{\mathrm{T}} \tag{2.10}$$

where L is a lower triangular matrix with leading diagonal elements being one, L^{T} is its transpose, and $D(\omega)$ is a diagonal matrix.

Based on the result of D, the following Rayleigh quotient[39, 40] is used to estimate the eigenvalue solution, toaccelerate its convergence:

$$\lambda = \frac{D^{\mathrm{T}}KD}{D^{\mathrm{T}}MD} \tag{2.11}$$

2.3.2　Inverse iteration for single eigenpairs

The Sturm theorem[28-44] are introduced to fix the eigen intervals for containing the single eigen value or coincident eigen values in the research, which was a widely used technique for evaluate the upper or lower bounds of eigenvalues of any order. Consequently, using the conventional FEM, this research applies the inverse iteration to solve single eigensolutions, and appliesthe subspace iterationto solve multiple eigensolutions, respectively. To solve the single eigenpairs, the inverse iteration technique is used in the following procedure:

$$\left.\begin{aligned} \overline{D}_{i+1} &= K_{\mathrm{a}}^{-1}MD_i \\ \mu_{i+1} &= \frac{\overline{D}_{i+1}^{\mathrm{T}}MD_i}{\overline{D}_{i+1}^{\mathrm{T}}M\overline{D}_{i+1}} \\ D_{i+1} &= \mathrm{sgn}(\mu_{i+1})\frac{\overline{D}_{i+1}}{\max(\overline{D}_{i+1})} \end{aligned}\right\} \quad i = 0, 1, \cdots \tag{2.12}$$

where i is the loop index.

The above inverse iteration procedure is terminated when the following conditions are met:

$$|\mu_{i+1} - \mu_i| < Tol \text{ and } \max|D_{i+1} - D_i| < Tol \tag{2.13}$$

2.3.3　Subspace iteration for coincident eigenpairs

In thecase of coincident eigenpairs, the eigenpairsare defined as follows:

$$\Lambda = \begin{bmatrix} \mu_1 & & & \\ & \mu_2 & & \\ & & \cdots & \\ & & & \mu_{N_{\mathrm{r}}} \end{bmatrix}, \quad D = [D_1, D_2, \cdots, D_{N_{\mathrm{r}}}] \tag{2.14}$$

where Λ and D are coincident eigenvalues and eigenfunctions, respectively, and $N_{\mathrm{r}}(>1)$ is the number of coincident eigenpairs.

The coincident eigenpairsare solved using the subspace iteration to find N_{r} eigenpairs simultaneously, withthe following procedure:

$$
\left.
\begin{aligned}
&\overline{\boldsymbol{D}}^{(i+1)} = \boldsymbol{K}_{\mathrm{a}}^{-1} \boldsymbol{M} \boldsymbol{D}^{(i)} \\
&\boldsymbol{K}_{\mathrm{r}}^{(i+1)} = \overline{\boldsymbol{D}}^{(i+1)\mathrm{T}} \boldsymbol{M} \boldsymbol{D}^{(i)}, \quad \boldsymbol{M}_{\mathrm{r}}^{(i+1)} = \overline{\boldsymbol{D}}^{(i+1)\mathrm{T}} \boldsymbol{M} \overline{\boldsymbol{D}}^{(i+1)} \\
&\mathrm{solve} \Rightarrow \boldsymbol{K}_{\mathrm{r}}^{(i+1)} \boldsymbol{V}^{(i+1)} = \boldsymbol{M}_{\mathrm{r}}^{(i+1)} \boldsymbol{V}^{(i+1)} \boldsymbol{\Lambda}^{(i+1)} \\
&\widetilde{\boldsymbol{D}}^{(i+1)} = \overline{\boldsymbol{D}}^{(i+1)} \boldsymbol{V}^{(i+1)} \\
&\boldsymbol{D}_k^{(i+1)} = \mathrm{sgn}(\mu_k^{(i+1)}) \frac{\widetilde{\boldsymbol{D}}_k^{(i+1)}}{\max\limits_j |\widetilde{D}_{k,j}^{(i+1)}|} \quad k=1, \cdots, N_{\mathrm{r}}
\end{aligned}
\right\} \quad i=0, 1, \cdots \qquad (2.15)
$$

This procedure ends when all the solutions of the N_{r} eigenpairs satisfy the following stop criteria:

$$
|\mu_k^{(i+1)} - \mu_k^{(i)}| < Tol \text{ and } \max |D_{k,j}^{(i+1)} - D_{k,j}^{(i)}| < Tol \qquad (2.16)
$$

After the above inverse or subspace iterations converge, the FE solutions are obtained. However, if the current mesh is not sufficiently fine, the accuracy of the FE solutions must be estimated by the more accurate superconvergent solution discussed in the following section.

2.4　Error estimation for eigenfunctions

The superconvergent patch recovery displacement method was developed[9, 45, 46] to acquire the superconvergent displacements of the FE solutions in static and dynamic problems. The displacements provided by this method can be applied to eigenfunctions. For example, if element e is a superconvergent computation element and elements $e-1$ and $e+1$ can be considered its neighbouring elements, all FE nodes in patched elements $e-1$, e, and $e+1$ are selected for the computation process. Further, the superconvergent displacements for element e can be computed as[9] :

$$
\boldsymbol{u}^*(x) = \sum_{i=1}^r \boldsymbol{N}_i(x) \boldsymbol{u}_i^h + \sum_{i=1}^s \boldsymbol{N}_i(x) \overline{\boldsymbol{u}}_i^* \qquad (2.17)
$$

where r $(=2)$ is the number of end nodes, s is the number of internal nodes and $\boldsymbol{N}_i$ is the shape function matrix. Using the high-order shape function interpolation, the shape function polynomial order is increased, $r+s > m+1$. To optimise the superconvergent order $O(h^{2m})$ for displacements at the end nodes, the displacement recovery field can be expressed for FE nodes as:

$$
\overline{u}_i^*(x) = \boldsymbol{P}\boldsymbol{a}, \quad i=1, \cdots, n_{\mathrm{d}} \qquad (2.18)
$$

where $\boldsymbol{P}$ is the given function vector and $\boldsymbol{a}$ can be obtained by the least-squares fitting technique for the coincidence of displacements at the end nodes in both the recovery and the conventional FE fields. The superconvergent displacements of the recovery field are used in Equation (2.17) to obtain the superconvergent solutions on element e. The vector coefficients $\boldsymbol{P}$ and $\boldsymbol{a}$ forms used for this study were:

$$\boldsymbol{P} = \begin{bmatrix} 1 & x & \cdots & x^p \end{bmatrix}, \ \boldsymbol{a} = \begin{bmatrix} a_1 & a_2 & \cdots & a_m \end{bmatrix}^T \tag{2.19}$$

The value of $\boldsymbol{a}$ was determined by obtaining the minimum value of the following functional, so that the product result on the positions of the FE nodes in Equation (2.19) is equal to the displacement values:

$$\Pi = \sum_{j=1}^{n} (u_i^*(x_j) - \boldsymbol{P}(x_j)\boldsymbol{a}), \ i = 1, \ \cdots, \ n_d \tag{2.20}$$

where n is the node number of all elements patched together.

Using the least square method to solve Equation (2.21), the value of the coefficient $\boldsymbol{a}$ can be obtained as:

$$\boldsymbol{a} = \boldsymbol{A}^{-1}\boldsymbol{b} \tag{2.21}$$

The coefficient matrices $\boldsymbol{A}$ and $\boldsymbol{b}$ are defined as follows:

$$\boldsymbol{A} = \sum_{j=1}^{n} \boldsymbol{P}(x_j)^T \boldsymbol{P}(x_j), \ \boldsymbol{b} = \sum_{j=1}^{n} \boldsymbol{P}(x_j)^T u_i^h(x_j), \ i = 1, \ \cdots, \ n_d \tag{2.22}$$

After the coefficient $\boldsymbol{a}$ is determined, the superconvergent solutions of the displacement of the piecewise elements can be obtained from Equation (2.17).

The estimated eigenvalue has a stationary value when taken over all possible functions that satisfy the essential boundary conditions. Stationary values are superconvergent eigenvalues. Further, the superconvergent solutions of the displacements can be used in the Rayleigh quotient[39, 40] to obtain the estimation of the eigenvalue from the following equation:

$$\lambda^* = \frac{a(\boldsymbol{u}^*, \boldsymbol{u}^*)}{b(\boldsymbol{u}^*, \boldsymbol{u}^*)} \tag{2.23}$$

where $a(\bullet)$ and $b(\bullet)$ are the strain and the kinematic energy inner products, respectively. To determine if the solution on the current mesh satisfy the given tolerance, the error isestimated by[9]:

$$\|\boldsymbol{e}^*\| \leqslant Tol \cdot [(\|\boldsymbol{u}^h\|^2 + \|\boldsymbol{e}^*\|^2)/n_e]^{1/2} \tag{2.24}$$

where n_e is the number of elements, and $\|\boldsymbol{e}^*\|$ is error in energy norm:

$$\|\boldsymbol{e}^*\| = [a(\boldsymbol{e}^*, \boldsymbol{e}^*)]^{1/2} = \left[\int_{\Omega} \boldsymbol{e}^{*T} \boldsymbol{L} \boldsymbol{e}^* \, dx\right]^{1/2} \tag{2.25}$$

where $\boldsymbol{e}^* = \boldsymbol{u}^* - \boldsymbol{u}^h$, and Ω is solution domain. Equation (2.24) can be equivalently rewritten in the following form:

$$\xi = \frac{\|\boldsymbol{e}^*\|}{\bar{e}} \text{ with } \bar{e} = Tol \cdot [(\|\boldsymbol{u}^h\|^2 + \|\boldsymbol{e}^*\|^2)/n_e]^{1/2} \tag{2.26}$$

where ξ should satisfy:

$$\xi \leqslant 1 \tag{2.27}$$

2.5　*h*-version mesh refinement

If Equation (2.27) is not satisfied, the corresponding element needs to be subdivided into uniform sub-elements by inserting some interior nodes through h-refinement[9] . These are computed by:

$$h_{\text{new}} = \xi^{-1/m} h_{\text{old}} \tag{2.28}$$

where h_{new} is the length of the sub-element and h_{old} is the original length of element e. The above element subdivision approach is implemented as follows:

$$n_{\text{new}} = \min(\lfloor \xi^{1/m} \rfloor, d) \tag{2.29}$$

where n_{new} is the number of sub-elements after element subdivision, the symbol $\lfloor \cdot \rfloor$ represents the "floor" operator (i. e. , the rounding up to the nearest integer), and d is the limit number needed to avoid too many redundant elements. Each element e that does not satisfy the pre-specified error toleranceis uniformly subdivided by h-version mesh refinement, e. g. , $h_{\text{new}} = h_{\text{old}}/6$, as shown in Figure 2.2.

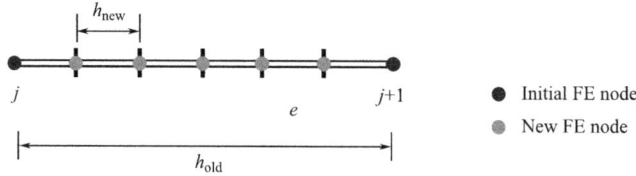

Figure 2.2　Diagram of uniform subdivision on element e
by h-version mesh refinement (e. g. $h_{\text{new}} = h_{\text{old}}/6$)

The non-uniform mesh in high-performance adaptive analysis process is the crucial factor to efficient computation, in which the minimum number of elements and optimized distribution of nodes are used to quickly reduce the solution errors and derive high-precision solutions[47-49] . In this research, based on the error estimation in energy norm, the solution error of one element can be estimated using the superconvergent solution for each element, and the error may be accurately reduced by using uniform subdivision for this element. Meanwhile, in order to avoid redundant element in such uniform subdivision, this proposed method limit the maximum number of subdivision in each adaptive stage. From the overall point of view, the mesh on the global domain are non-uniform because each element is independent for error estimation and mesh subdivision. With the subdivision and increase of elements, the non-uniform tendency of elements become prominent to suit the variation of eigenfunctions. Through the numerical examples in this chapter andsome verification examples in previous researches[9] , this error estimation in energy norm and uniform subdivision on each element for mesh refinement showgood accuracy, reliability and effectiveness for eigensolutions.

2.6　Numerical examples

By proposing some novel techniques, such as the maximum number of subdivision in eachadaptive stage, the computation schemes for coincident eigenvalues, and the variable matrices and adjacent approximate eigenvalues in vector Sturm-Liouville problems, the superconvergent patch recovery method is developed for the eigensolutions of continues order for eigenproblems in system ofsecond order ODEs. This section presents sometypical and representative numerical examples showing the performance of the proposed method and algorithm, which were implemented in a programme written in Fortran 90. The programme was run on a DELL Optiplex 380 Intel (R) Core (TM) 2. 93 GHz desktop computer. The pre-specified error tolerance Tol was set to 10^{-4} and the degree $m = 4$ for each element.

Tomeasure the accuracy of results, the error ε_λ between the exact eigenvalue λ and the computed eigenvalue λ^h was defined as:

$$\varepsilon_\lambda = \frac{|\lambda - \lambda^h|}{1 + |\lambda|} \tag{2.30}$$

When exact solutions are not available, some high-precision numerical solutions will be used and ananalogous error ε_λ^* is defined as follows:

$$\varepsilon_\lambda^* = \frac{|\bar{\lambda} - \lambda^h|}{1 + |\bar{\lambda}|} \tag{2.31}$$

2. 6. 1　Example 1: Coefficients of variable matrices and adjacent approximate eigenvalues

To illustrate the feasibility and accuracy of the proposed method for SL problems with coefficients of variable matrices and adjacent approximate eigenvalues, a typical example with exact solutions was carefully chosen. Consider the following vector SL problems with two unknown function components $\boldsymbol{u} = (u_1, u_2)^\mathrm{T}$ in which the coefficients of variable matrices are as follows[23]:

$$\boldsymbol{A} = \boldsymbol{R} = \boldsymbol{I}, \ \boldsymbol{C} = \frac{x}{2} \begin{pmatrix} 3 & -1 \\ -1 & 3 \end{pmatrix} \tag{2.32}$$

where $\boldsymbol{B} = \boldsymbol{0}$; $\boldsymbol{I}$ is theidentity matrix; the solution domain is $0 \leqslant x \leqslant 1$; and the boundary conditions are fixed boundaries on both ends.

In this example, there are some solutions for two adjacent eigenvalues, rather than coincident eigenvalues, but it is difficult to establish the robustness of the algorithm. Using the proposed method, the first ten eigenpairs were computed as listed in Table 2. 1. To display the trend of adjacent approximate eigenvalues, all the eigenvalues were plotted in Figure 2. 3. It can be seen that each group of two adjacent eigenvalues are quite approximate, and the adjacent eigenvalues are distinguished as single eigenvalues. These single eigenvalues are appropriately computed by the proposed adaptive inverse interaction and the errors of the solutions satisfy the pre-specified error tolerance of 10^{-4} when compared to the exact eigenvalues as shown in Table 2. 1. In order to verify the effectiveness of the proposed method for solving high-precision solutions, the solutions under the stricter error toleranceof 10^{-8} are computed; the computed eigenvalues of first three orders are10. 36850731, 10. 86521581, and 39. 97874485, and corresponding errors ε_λ

are 9.68×10^{-9}, 9.27×10^{-9}, and 1.22×10^{-9}, respectively. It can be seen that the satisfactory solutions are successfully obtained. Other methods have been developed for satisfying the requirements of the solutions accuracy needed to distinguish between these adjacent approximate eigenvalues. The functional-discrete method[23] is adapted to obtain the first ten solutions listed in Table 2.1. As can be seen, these results are high-precision; however, the eigenfunctions corresponding to these adjacent approximate eigenvalues are not provided, and it is unclear whether they can be properly solved.

Computed eigenvalues and errors for Example 1, coefficients of variable

matrices and adjacent approximate eigenvalues for vector *SL* problems   Table 2.1

k	Proposed method		Functional discrete method[1]		Exactvalues[1]
	λ_k^h	ε_λ	$\bar{\lambda}_k$	$\bar{\varepsilon}_\lambda$	λ_k
1	10.36827	2.09×10^{-5}	10.3685071396	1.95×10^{-9}	10.3685072
2	10.86594	6.10×10^{-5}	10.8652153552	2.99×10^{-8}	10.8652157
3	39.97817	1.40×10^{-5}	39.9787448114	5.25×10^{-10}	39.9787448
4	40.47974	3.35×10^{-7}	40.4797264325	8.29×10^{-9}	40.4797261
5	89.32653	1.16×10^{-6}	89.3266345432	7.64×10^{-12}	89.3266345
6	89.82753	3.42×10^{-6}	89.8272193326	4.08×10^{-12}	89.8272193
7	158.4181	2.70×10^{-5}	158.4137898143	1.69×10^{-13}	158.413790
8	158.9129	7.80×10^{-6}	158.9141480051	2.76×10^{-12}	158.914148
9	247.2417	6.09×10^{-6}	247.2401893286	2.42×10^{-14}	247.240189
10	247.7438	1.36×10^{-5}	247.7404272325	3.38×10^{-13}	247.740427

Source: ① Results from paper [23].

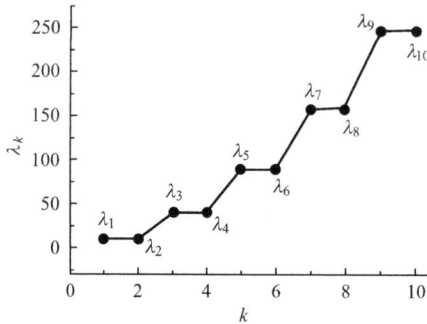

Figure 2.3   Adjacent approximate eigenvalues of Example 1, coefficients of variable
matrices and adjacent approximate eigenvalues of vector *SL* problems

Using the proposed adaptive inverse interaction for single eigenpairs, the eigenfunctions of the adjacent approximate eigenvalues are solved; Figure 2.4 shows the selected eigenfunctions and corresponding final meshes for 1st and 2nd, 9th and 10th orders. These eigenfunctions have symmetric and antisymmetric characteristics for the two function components (u_1, u_2). The eigenfunctions (u_{11}^h, u_{12}^h) and (u_{91}^h, u_{92}^h) are affiliated to the first adjacent approximate eigenvalues, which are symmetric (Figures 2. 4 (a) and 2. 4 (c)), and the eigenfunctions (u_{21}^h, u_{22}^h) and (u_{101}^h, u_{102}^h) are affiliated to the second one, which are antisymmetric (Figures 2. 4 (b) and 2. 4 (d)). Further, these curves show a gradual change trend and, therefore, a reasonable uniform distributed mesh in the global domain was generated as shown in the horizontal axis.

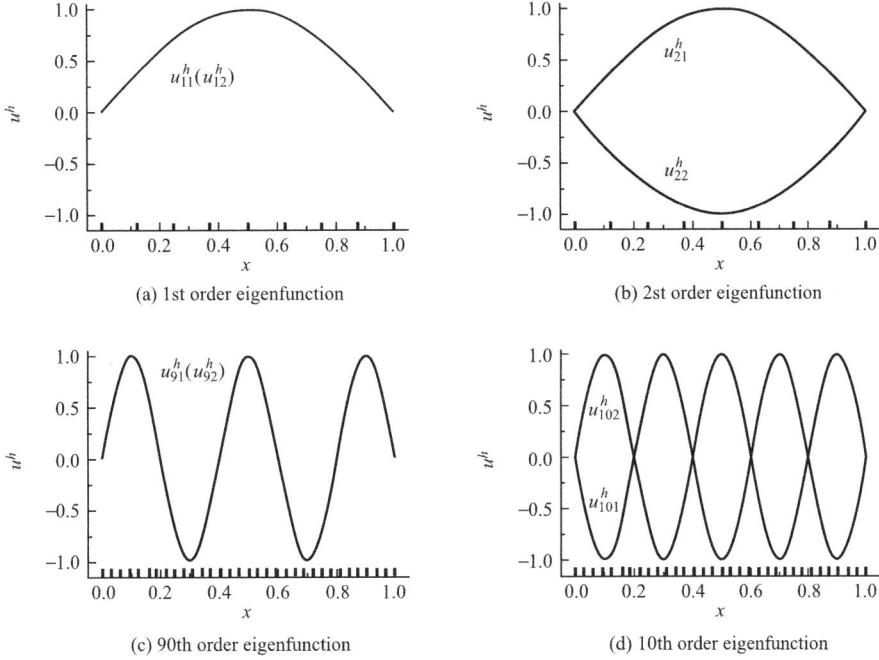

(a) 1st order eigenfunction

(b) 2st order eigenfunction

(c) 90th order eigenfunction

(d) 10th order eigenfunction

Figure 2. 4　Computed symmetric and antisymmetric eigenfunctions for 1st and 2nd, 9th and 10th orders and corresponding final meshes of Example 1; coefficients of variable matrices and adjacent approximate eigenvalues of vector SL problems

2. 6. 2　Example 2: Benchmark eigenproblems in SL12F

To test the validity of the proposed algorithm for the vector SL problems, a challenging benchmark example with four unknown function components, $\boldsymbol{u} = (u_1, u_2, u_3, u_4)^{\mathrm{T}}$, and coefficients of variable matricesis chosen as follows:

$$\boldsymbol{A} = \boldsymbol{R} = \boldsymbol{I}, \ C_{ij} = \frac{1}{\max(i, j)}\cos(x) + \frac{\delta_{ij}}{x^i}, \ i, j = 1, \cdots, 4 \qquad (2.33)$$

where $\boldsymbol{B} = \boldsymbol{0}$; δ_{ij} is the Kronecker Delta function; the solution domain is $0 \leqslant x \leqslant 1$; and the boundary conditions are fixed boundarieson both ends.

SL12F[24] used this example as a benchmark case to detect the effectiveness for acquiring the high-precision solutions of eigenvalues. Due to the complexity of this problem, its exact solutions are not available. Instead, the error ε_λ^* is analysed using the high-precision numerical solutions computed by SL12F. The numerical solutions were computed by setting the strict error tolerance $Tol = 10^{-12}$ for more exact decimal digits. The selected first ten eigenvalues ($k = 1 \sim 10$), errors ε_λ^* and final number of elements computed are listed in Table 2. 2. The errors of the eigenvalues satisfy the pre-specified error tolerance of 10^{-4}, which confirms the accuracy of the solutions computed through the proposed method.

Computed results for eigenvalues and errors of Example 2, benchmark eigenproblems in SL12F

Table 2. 2

k	Proposed method			SL12F[①]
	λ_k^h	ε_λ^*	Final number of elements	$\bar{\lambda}_k$
1	14. 94173	4.43×10^{-6}	13	14. 9418005
2	17. 04384	1.90×10^{-5}	17	17. 0434966
3	21. 38025	7.62×10^{-6}	22	21. 3804205
4	26. 92052	7.57×10^{-6}	22	26. 9207313
5	51. 82594	4.41×10^{-6}	22	51. 8257072
6	55. 80365	2.36×10^{-6}	24	55. 8035161
7	67. 49423	2.85×10^{-6}	24	67. 4944253
8	84. 92138	4.15×10^{-6}	24	84. 9210235
9	112. 8618	2.24×10^{-5}	30	112. 864353
10	117. 9062	4.97×10^{-6}	30	117. 905610

Source: ① Results from paper[24].

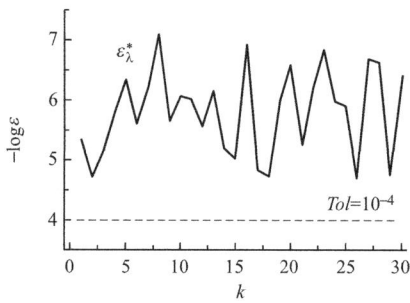

Figure 2. 5　Errors in the computed first thirty eigenvalues of Example 2

To show the validity of the computation for extensive continuous orders, the errors of the first thirty eigenvalues are shown in Figure 2. 5. As can be seen, the accuracy is much lower than the initial pre-specified error tolerance $Tol = 10^{-4}$ because the eigenvalue computed by the Rayleigh quotient, based on the high-precision solution of eigenfuction, has the higher-order accuracy. This shows the reliability of this method to control the error of the eigenfunction and thus use it as the stop criterion.

Because reliable solutions of eigenfunctions of vector SL problems are challenging, most of best-established numerical methods and models, including SL12F, cannot provide these. The proposed method addresses this issue for the first three order vector functions and their corresponding final meshes on the horizontal axes are shown in Figures 2. 6 (a), 2. 6 (b), and 2. 6 (c), respectively. The vibration mode is affected by the fixed boundary conditions and computed values of all the mode components (u_{31}^h, u_{32}^h, u_{33}^h, u_{34}^h) at the fixed boundaries are 0. As the order of eigenvalues increases, the eigenfunctions become complex and require more elements and denser meshes. Observing the marked meshes on the horizontal axis, it is evident that the adaptive analysis process automatically and properly arranges more elements for the sharply varied parts of these eigenfunctions and that the non-uniform meshes are generated for each order.

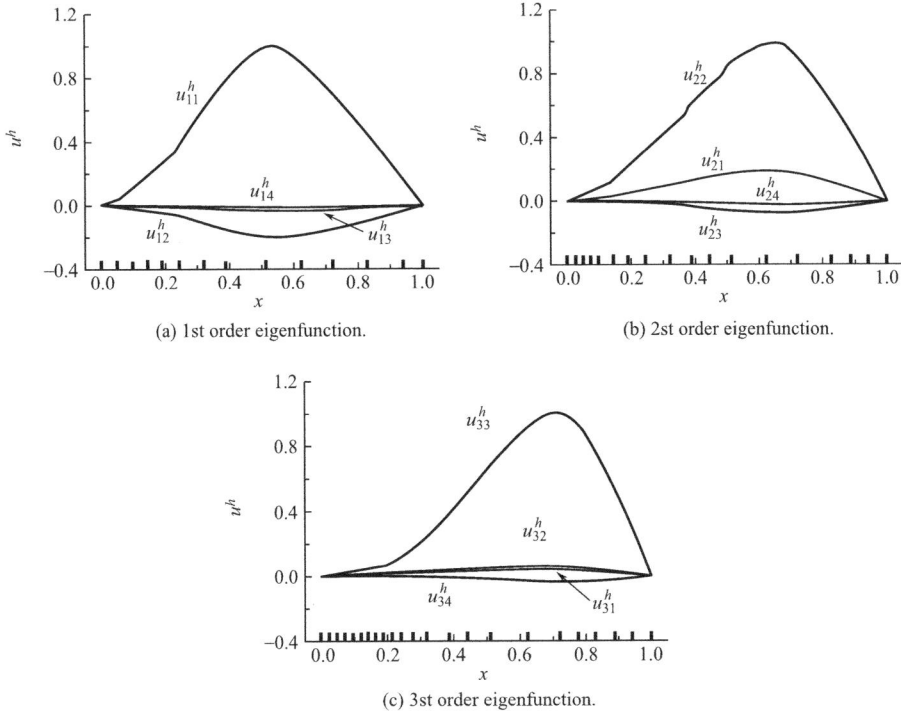

(a) 1st order eigenfunction.

(b) 2st order eigenfunction.

(c) 3st order eigenfunction.

Figure 2. 6　Computed first three eigenfunctions and corresponding final
meshes of Example 2, benchmark eigenproblems in SL12F

2. 6. 3　Example 3: Coefficients of constant matrices and coincident eigenvalues

To illustrate the feasibility and accuracy of the proposed method for coincident eigenvalues, an example with multiple coincident eigenvalues was chosen, so that the computed results obtained from the proposed method could be compared with the exact solutions. Consider the following vector SL problems with three unknown function components $\boldsymbol{u} = (u_1, u_2, u_3)^{\mathrm{T}}$ in which the coefficients of constant matrices are as follows[22]:

$$\boldsymbol{A} = \begin{pmatrix} 11 & 6 & 3 \\ 6 & 12 & 2 \\ 2 & 2 & 1 \end{pmatrix}, \ \boldsymbol{C} = 0, \ \boldsymbol{R} = \begin{pmatrix} 38 & 24 & 12 \\ 24 & 18 & 8 \\ 12 & 8 & 4 \end{pmatrix} \tag{2.34}$$

where $\boldsymbol{B} = \boldsymbol{0}$; the solution domain is $0 \leqslant x \leqslant 1$; and the boundary conditions are fixed boundarieson both ends.

The shooting method[22] has been adapted to obtain the approximate coincident eigenvalues solution. The solution $\overline{\lambda}_{k1}$ is analysed beginning with random real guess values; $\overline{\lambda}_{k2}$ is analysed beginning with integer guess values, which could improve the accuracy of the solutions by continued bisection. Unfortunately, the method does not control the errors of the solutions of coincident eigenvalues, as listed in Table 2.3. By the proposed method, the first ten eigenpairs are also computed. In the computation process, the coincident eigenvalues (such as 2nd and 3rd orders, 5th and 7th orders, and 9th and 10th orders) are analysed using the adaptive subspace interaction. The errors of all coincident eigenvalues computed by the proposed method satisfy the pre-specified error tolerance when compared with the exact eigenvalues.

Computed results for eigenvalues and errors for Example 3, coefficients of constant matrices and coincident eigenvalues for the vector *SL* problems   Table 2.3

k	Proposed method		Shooting method[1]		Exactvalues[1]
	λ_k^h	ε_λ	$\bar{\lambda}_{k1}$	$\bar{\lambda}_{k2}$	λ_k
1	0.250003	2.40×10^{-6}	0.250479	0.250488	0.25
2	1.000025	1.25×10^{-5}	1.000479	0.999512	1.00
3	1.000025	1.25×10^{-5}	1.000479	1.000488	1.00
4	2.250083	2.55×10^{-5}	2.249991	2.250977	2.25
5	4.000047	9.40×10^{-6}	3.999014	3.998047	4.00
6	4.000047	9.40×10^{-6}	3.999014	4.001953	4.00
7	4.000047	9.40×10^{-6}	3.999014	4.001953	4.00
8	6.250033	4.55×10^{-6}	6.249014	6.251953	6.25
9	9.000052	5.20×10^{-6}	8.997061	8.996094	9.00
10	9.000052	5.20×10^{-6}	8.997061	9.003906	9.00

Source: [1]Results from paper[22].

To evaluate the effectiveness of the proposed method to solve the eigenfunctions corresponding to the coincident eigenvalues, the computed eigenfunctions for 5th, 6th, and 7th orders and their corresponding final meshes are shown in Figures. 7 (a)～(b), Figures 2.7 (c)～(d), and Figures 2.7 (e)～(f), respectively. The eigenfunction is affected by the fixed boundary conditions, and values of all components (u_{31}^h, u_{32}^h, u_{33}^h) in these eigenfunctions at the fixed boundaries are 0. As can be seen, using the adaptive subspace iteration, these eigenfunctions are solved at the same time under identical meshes. Further, the mesh is adjusted and optimised according to the eigenfunction change. The dense mesh is used in the domain with dramatic eigenfunction change, and the sparse mesh is used in the domain with gradual eigenfunction change. Finally, a non-uniform mesh is generated.

2.6.4 Example 4: Free vibration of planar elliptic beams

In this example, the typical applications of the proposed method on free vibration problems of planarcurved beams will be presented. The geometric model of a planar elliptic beam is shown in Figure 2.8, where a and b are the major and minor axis, and θ is the angle of thisplanar elliptic beams.

(a) 5th order eigenfunction(u_{51}^h, u_{52}^h).

(b) 5th order eigenfunction(u_{53}^h).

(c) 6th order eigenfunction(u_{61}^h, u_{62}^h).

(d) 6th order eigenfunction(u_{63}^h).

(e) 7th order eigenfunction(u_{71}^h, u_{72}^h).

(f) 7th order eigenfunction(u_{73}^h).

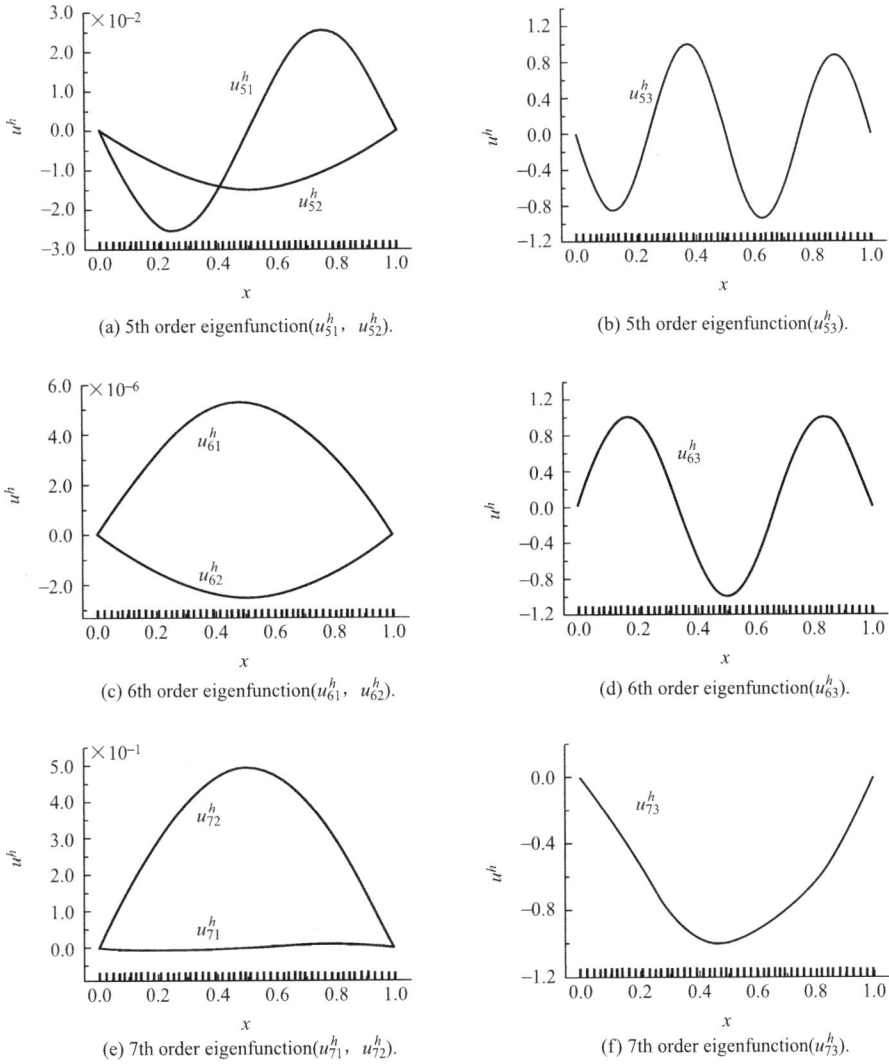

Figure 2. 7 Computed three eigenfunctions corresponding to the coincident eigenvalues for 5th-7th orders and corresponding final meshes of Example 3, coefficients of constant matrices and coincident eigenvalues of vector SL problems

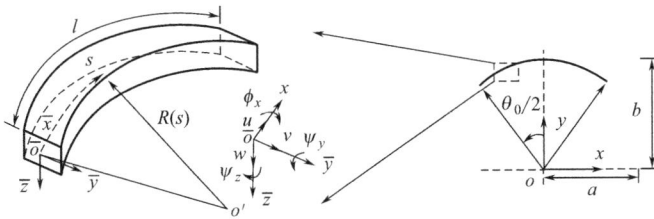

Figure 2. 8 Geometric model of the planar elliptic beam used in Example 4

The in-plane free vibration problems of planar elliptic beams are eigenproblems in systems of second order ODEs with frequency ω and three unknown displacement components $\boldsymbol{u} = (u_1, u_2, u_3)^{\mathrm{T}} = \{u, v, \psi_z\}^{\mathrm{T}}$ in plane xOy. For this problem, the coefficients in the matrix form are as follows[17]:

$$
\boldsymbol{A} = \begin{pmatrix} EA & 0 & 0 \\ 0 & \kappa GA & 0 \\ 0 & 0 & EI \end{pmatrix}, \quad \boldsymbol{B} = \begin{pmatrix} 0 & -k_s EA & 0 \\ k_s \kappa GA & 0 & -\kappa GA \\ 0 & 0 & 0 \end{pmatrix}
$$

$$
\boldsymbol{C} = \begin{pmatrix} k_s^2 \kappa GA & 0 & -k_s \kappa GA \\ 0 & k_s^2 EA & 0 \\ -k_s \kappa GA & 0 & \kappa GA \end{pmatrix}, \quad \boldsymbol{R} = \begin{pmatrix} \rho A & 0 & 0 \\ 0 & \rho A & 0 \\ 0 & 0 & \rho I \end{pmatrix}
$$

$$(2.35)$$

where A, I, κ and ρ are the cross-sectional area, moment of inertia, correction coefficient of shear stiffness, and the material density, respectively; E and G are the elastic modulus and shear modulus; R and $k_s (= 1/R)$ are the curvature radius and the responding curvature.

This planar elliptic beam has the following elliptic equation:

$$
\frac{x^2}{a^2} + \frac{y^2}{b^2} = 1, \quad a > b > 1 \tag{2.36}
$$

The curvature of elliptic curved beam varies as follows:

$$
k_s = \frac{|y''|}{[1 + (y')^2]^{\frac{3}{2}}} \tag{2.37}
$$

where k_s varies with x, increasing the difficulty of finding a solution; the solution domain is $0 \leqslant x \leqslant 1$.

The basic physical parameters of this planar elliptic are:

$$
A = 0.004 \text{ m}, \ E = 70 \text{ GPa}, \ \kappa = 0.85, \ \kappa G/E = 0.3
$$
$$
\rho = 2777 \text{ kg/m}^3, \ I_z = 0.0001 \text{ m}^4 \tag{2.38}
$$

Continuous orders for frequencies and modes of curved beams in different kinds of geometries (ratios b/a and angles θ_0) and for various boundary conditions (such as clamped-clamped, clamped-hinged, and hinged-hinged) have been analysed to detect the accuracy and reliability of the proposed method. The first three eigenpairs are computed, and the numerical natural frequencies are transformed into non-dimensional values as $\tilde{\omega} = \omega l^2 \sqrt{\rho A/EI}$ for simplicity (Table 2.4). The dynamic stiffness method[17] has been introduced to obtain approximate solutions. Comparing the results of the above two methods, it is evident that they are in agreement, which highlights the accuracy and reliability of the proposed method.

Computed results for frequencies of Example 4, free vibration of planar elliptic beams

Table 2.4

b/a	θ_0	k	Clamped-clamped		Clamped-hinged		Hinged-hinged	
			Proposed method	Dynamic stiffness method[①]	Proposed method	Dynamic stiffness method[①]	Proposed method	Dynamic stiffness method[①]
			$\widetilde{\omega}_k^h$	$\widetilde{\omega}_k$	$\widetilde{\omega}_k^h$	$\widetilde{\omega}_k$	$\widetilde{\omega}_k^h$	$\widetilde{\omega}_k$
0.2	60°	1	92.97536	93.1679	70.50068	70.5699	54.91535	54.9307
		2	227.4738	228.641	188.4397	189.095	151.7512	152.053
		3	425.5924	428.962	377.4245	379.686	330.1734	331.532
	120°	1	49.06749	49.0833	47.78234	47.9755	47.67364	47.6857
		2	77.72862	77.8741	63.28713	63.3659	50.08538	50.1204
		3	154.5412	154.973	136.2353	136.505	119.6977	119.854
	180°	1	43.86628	43.9775	38.55956	38.5975	33.51056	33.5267
		2	53.79595	53.8656	49.30078	49.3955	39.47331	39.5233
		3	122.7464	122.959	120.8838	121.096	120.8656	121.078
0.5	60°	1	120.3227	120.446	107.1543	107.198	101.7318	101.751
		2	220.6755	221.794	182.6026	183.223	146.3662	146.652
		3	418.7284	421.938	371.1451	373.301	324.5295	325.821
	120°	1	67.43523	67.5534	54.11178	54.1738	42.01836	42.0451
		2	84.09314	84.2185	81.64115	81.7583	75.95545	76.0361
		3	160.4025	160.647	150.7322	150.869	146.4912	146.575
	180°	1	35.94875	35.9884	27.37087	27.3890	19.89473	19.9020
		2	50.41136	50.5204	45.46445	45.5332	38.50727	38.5404
		3	109.9644	110.291	96.90023	97.1136	84.67194	84.7922
0.8	60°	1	155.4975	155.586	147.2627	147.351	137.4468	137.706
		2	209.2023	210.235	174.1112	174.642	148.8554	148.891
		3	406.6476	409.609	360.2089	362.174	314.8278	316.002
	120°	1	54.41390	54.4974	42.95123	42.9955	32.76831	32.7867
		2	92.09564	92.2983	83.63068	83.7741	73.78976	73.8734
		3	173.5345	173.668	161.2712	161.672	145.9642	146.255
	180°	1	22.85531	22.8741	17.13685	17.1453	12.05678	12.0598
		2	43.48537	43.5458	37.86814	37.9061	32.08254	32.1018
		3	82.95367	83.1160	73.93398	74.0436	65.45792	65.5254

Source: ① Results from paper[17].

To assess the effects of the boundary conditions on frequencies, the computed first five eigenvalues of a planar elliptic beam with $b/a = 0.2$, $\theta_0 = 60°$ are listed in Figure 2.9. As can be seen, as the order increases the frequency value increases. Moreover, the frequency magnitude is affected by the boundary conditions. The more restrictive the

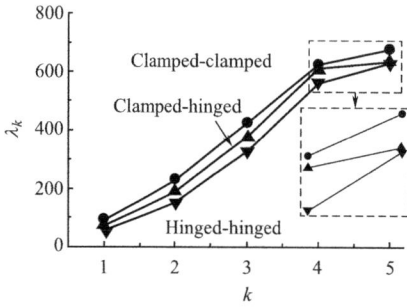

Figure 2.9 The computed five three eigenvalues for different boundary conditions of Example 4; free vibration of planar elliptic beam with $b/a = 0.2$, $\theta_0 = 60°$

boundary, the higher the frequency value, reaching a maximum when the beam is clamped at both ends (clamped-clamped boundary condition). The second highest value occurs when the two ends are fixed and simply supported, respectively (clamped-hinged), and the lowest when the two ends are simply supported (hinged-hinged).

To assess the effects of the boundary conditions on modes and meshes, the 3rd order vibration modes and their corresponding final meshes are computed for a planar elliptic beam with $b/a = 0.8$ and $\theta_0 = 60°$ (Figure 2.10).

Actually, in computation procedure for eigensolutions of one order, the frequency λ and the associated n_d dimensional vector vibration modes $\boldsymbol{u}(x) = (u_1(x), \cdots, u_{n_d}(x))^T$ in systems of second order ODEs are solved together on one set of mesh; the error estimation and mesh refinement are implemented in an identical adaptive stage. As a result, this set of mesh should be sui Table 2. to all vibration mode components of the current order. The vibration mode is affected by the boundary conditions: the values of the mode components (u_{31}^h, u_{32}^h, u_{33}^h) at the fixed boundary are 0, (u_{31}^h, u_{32}^h) at the simply supported boundary is 0, and u_{33}^h is not 0. The distribution of the mesh that is needed is also affected by the boundary constraints. Due to these strong constraints, the vibration modes under the fixed ends have dramatic changes in the global domain, requiring a relatively denser mesh, as shown in Figures 2.10 (a)~(d) (final number of elements = 32). However, due to weak boundary constraint, the vibration modes of the simply supported ends change gradually in the global domain, requiring a relatively sparser mesh, as shown in Figures 2.10 (e)~(f) (final number of elements = 16). As can be seen, the adaptive analysis process automatically and properly arranges more elements for the sharply varied parts of these modes and non-uniform meshes are generated.

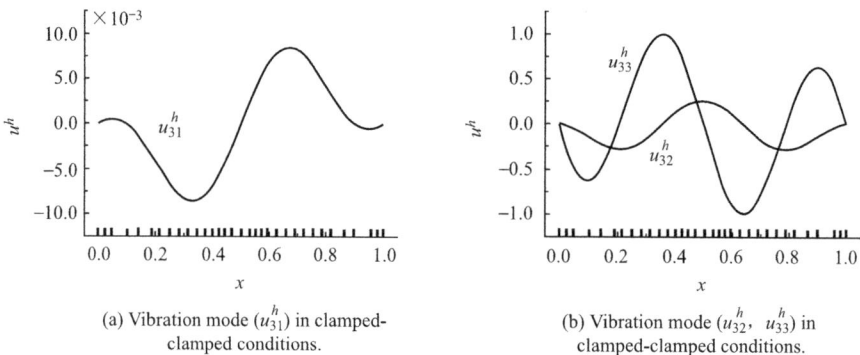

(a) Vibration mode (u_{31}^h) in clamped-clamped conditions.

(b) Vibration mode (u_{32}^h, u_{33}^h) in clamped-clamped conditions.

Figure 2.10 The computed 3rd order vibration modes and corresponding final meshes of Example 4, free vibration of planar elliptic beam with $b/a = 0.8$, $\theta_0 = 60°$ (one)

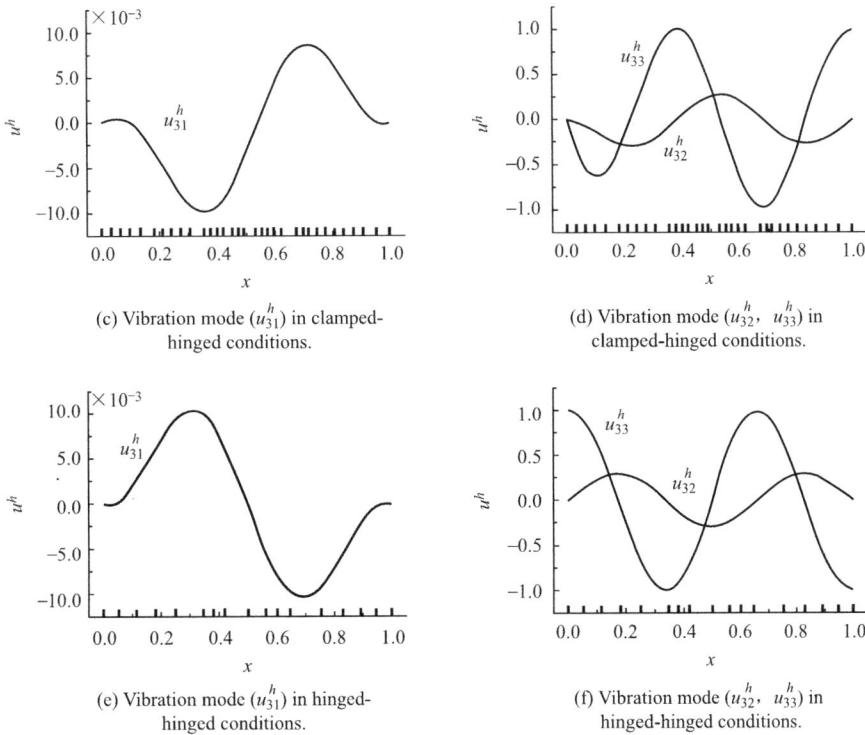

(c) Vibration mode (u_{31}^h) in clamped-hinged conditions.

(d) Vibration mode (u_{32}^h, u_{33}^h) in clamped-hinged conditions.

(e) Vibration mode (u_{31}^h) in hinged-hinged conditions.

(f) Vibration mode (u_{32}^h, u_{33}^h) in hinged-hinged conditions.

Figure 2.10   The computed 3rd order vibration modes and corresponding final meshes of Example 4, free vibration of planar elliptic beam with $b/a = 0.8$, $\theta_0 = 60°$ (two)

2.7   Conclusions

In this study, a novel h-version adaptive FE procedure for accurate computation of the eigenvalues and vector eigenfunctions for general eigenproblems in systems of second order ODEs was proposed. This procedure establishes non-uniform refined meshes to suit the change of eigenfunctions and computed solutions that satisfy the pre-specified error tolerances. Numerical examples are used to evaluate the effectiveness and reliability of the method. The conclusions are as follows:

(1) The superconvergent patch recovery displacement method and high-order shape function interpolation technique are introduced to obtain the superconvergent solution of eigenfunctions and eigenvalues using the Rayleigh quotient. Error estimates for eigenfunctions and h-version mesh refinement were proposed; further, non-uniform refined meshes are established to suit the change of the eigenfunctions, and the results satisfy the pre-specified error tolerance.

(2) Vector SL problems that present challenging issues includingthe coefficients of variable matrices, coincident and adjacent approximate eigenvalues, and continuous orders of eigenpairs, were analysed. Using the proposed method, the errors of computed

eigenvalues satisfy the pre-specified error tolerance, which verifies the accuracy of the solutions. Thus, the proposed method provides reliable solutions of eigenfunctions compared with other solvers for vector SL problems. The errors of computed eigenvalues are much lower than the initially pre-specified error tolerance because the eigenvalue is computed using the Rayleigh quotient. This quotient is based on the high-precision solution of the eigenfunction, which has a higher-order accuracy. This shows the reliability of controlling the errors of an eigenfunction and using it as the stop criterion. The adaptive analysis processes automatically and properly arrange more elements for the sharply varied parts of eigenfunctions, and non-uniform meshes are generated.

(3) Free vibration of curved beams under various geometries and boundary conditions were analysed. The solutions of continuous orders of frequencies and modes of a typical planar elliptic beam are consistent with the high-precision solutions obtained by using the dynamic stiffness method. The vibration modes under fixed boundary conditions show dramatic changes attributed to the strong boundary constraints. In these domains, the proposed mesh refinement procedure provides a relatively denser mesh.

(4) An adaptive subspace iteration was used to compute coincident eigenvalues and eigenfunctions under identical meshes; the errors of the coincident results satisfy the prespecified error toleranceas compared to the exact solutions. The mesh is adjusted and optimised according to the change of all these eigenfunctions. For adjacent approximate eigenvalues, rather than coincident eigenvalues, the adaptive inverse iteration method is employed to solve these eigenvalues separately, and anon-uniform refined mesh is established for each order to accommodate for the eigenfunction change.

This study is mainly focused on the h-version adaptive FEM on eigenproblems in system of second order ODEs. As shown in this chapter, it is important to solve the free vibration problems of curved beams and provide high-precision solutions satisfying the pre-specified error tolerance. Based on the high-precision solution of a natural frequency or vibration mode, the accurate identification for number, size and location of damages in structures may be achieved. Follow-up research may focus on extending the h-version adaptive FEM to detect multiple damages in structures.

References

[1] Akulenko L D, Gavrikov A A, Nesterov, et al. Numerical solution of vector Sturm-Liouville problems with Dirichlet conditions and nonlinear dependence on the spectral parameter [J]. Computational Mathematics and Mathematical Physics, 2017, 57 (9): 1484-1497.

[2] Su H, Banerjee. Development of dynamic stiffness method for free vibration of functionally graded Timoshenko beams [J]. Computers and Structures, 2015, 147: 107-116.

[3] Chelkak D, Korotyaev. Weyl-Titchmarsh functions of vector-valued Sturm-Liouville operators on the unit interval [J]. Journal of Functional Analysis, 2009, 257 (5): 1546-1588.

[4] Malamud M M. On the completeness of the system of root vectors of the Sturm-Liouville operator with general boundary conditions [J]. Functional Analysis and Its Applications, 2008, 42 (3): 198-204.

[5] Al-Gwaiz M A. Sturm-Liouville theory and its applications [J]. Berlin, Springer, 2008.

[6] Kang B, Riedel C H, Tan, et al. Free vibration analysis of planar curved beams by wave propagation [J]. Journal of Sound and Vibration, 2003, 260(1): 19-44.

[7] Sivadas K R, Ganesan. Free vibration and material damping analysis of moderately thick circular cylindrical shells [J]. Journal of Sound and Vibration, 1994, 172: 47-61.

[8] Chang K C, Kim. Modal-parameter identification and vibration-based damage detection of a damaged steel truss bridge [J]. Engineering Structures, 2016, 122, 156-173.

[9] Wang Y, Ju Y, Zhuang Z, et al. Adaptive finite element analysis for damage detection of non-uniform Euler-Bernoulli beams with multiple cracks based on natural frequencies [J]. Engineering Computations, 2018, 35 (3): 1203-1229.

[10] Fan W, Qiao. Vibration-based damage identification methods: a review and comparative study [J]. Structural Health Monitoring, 2011, 10 (1): 83-111.

[11] Yang Y, Yang. State-of-the-art review on modal identification and damage detection of bridges by moving test vehicles [J]. International Journal of Structural Stability and Dynamics, 2018, 18 (02): 1850025.

[12] Huang Y, Chen J, Luo, et al. A simple approach for determining the eigenvalues of the fourth-order Sturm-Liouville problem with variable coefficients [J]. Applied Mathematics Letters, 2013, 26 (7): 729-734.

[13] Shen C L, Shieh. On the multiplicity of eigenvalues of a vectorial Sturm-Liouville differential equation and some related spectral problems [J]. Proceedings of The American Mathematical Society, 1999, 127 (10): 2943-2952.

[14] Lee J. Free vibration analysis of delaminated composite beams [J]. Computers and Structures, 2000, 74 (2): 121-129.

[15] Jin G, Ye T, Ma X, et al. A unified approach for the vibration analysis of moderately thick composite laminated cylindrical shells with arbitrary boundary conditions [J]. International Journal of Mechanical Sciences, 2013, 75, 357-376.

[16] Raveendranath P, Singh G, Pradhan, et al. Free vibration of arches using a curved beam element based on a coupled polynomial displacement field [J]. Computers and Structures, 2000, 78 (4): 583-590.

[17] Tseng Y P, Huang C S, Lin, et al. Dynamic stiffness analysis for in-plane vibrations of arches with variable curvature [J]. Journal of Sound and Vibration, 1997, 207, 15-31.

[18] Zhou D, Cheung. The free vibration of a type of tapered beams [J]. Computer Methods in Applied Mechanics and Engineering, 2000, 188 (1): 203-216.

[19] Prikazchikov V G, Loseva. High-accuracy finite-element method for the Sturm-Liouville problem [J]. Cybernetics and Systems Analysis, 2004, 40 (1): 1-6.

[20] Andrew A L. Asymptotic correction of more Sturm-Liouville eigenvalue estimates [J]. Bit Numerical Mathematics, 2003, 43(3): 485-503.

[21] Marletta M. Automatic solution of regular and singular vector Sturm-Liouville problems [J]. Numerical Algorithms, 1993, 4 (1): 65-99.

[22] Dwyer H I, Zettl. Eigenvalue computations for regular matrix Sturm-Liouville problems [J]. Electronic Journal of Differential Equations, 1995, 1995 (5): 1-13.

[23] Bandyrskiĭ Bĭ, Gavrilyuk I P, Lazurchak I I, et al. Functional-discrete method (FD-method) for matrix Sturm-Liouville problems [J]. Computational Methods in Applied Mathematics, 2005, 5 (4): 362-386.

[24] Pryce J D. Numerical solution of Sturm-Liouville problems [J]. Clarendon Press: Oxford. , 1993,

[25] NAG Fortran library manual. Numerical Algorithms Group Ltd: Oxford. [Z]. 1999.

[26] Zienkiewicz O C, Taylor R L, Zhu, et al. The finite element method: its basis and fundamentals (7th Edition) [M]. Oxford UK: Elsevier (Singapore), 2015.

[27] Ascher U, Christiansen J, Russell, et al. Algorithm 569: COLSYS: Collocation software for boundary-value ODEs [J]. ACM Transactions on Mathematical Software, 1981, 7 (2): 223-229.

[28] Estep D, Ginting V, Tavener, et al. A posteriori analysis of a multirate numerical method for ordinary differential equations [J]. Computer Methods in Applied Mechanics and Engineering, 2012, 223, 10-27.

[29] Johansson A, Chaudhry J H, Carey V, et al. Adaptive finite element solution of multiscale PDE-ODE systems [J]. Computer Methods in Applied Mechanics and Engineering, 2015, 287, 150-171.

[30] Bieniasz L K. Adaptive solution of BVPs in singularly perturbed second-order ODEs, by the extended Numerov method combined with an iterative local grid h-refinement [J]. Applied Mathematics and Computation, 2008, 198 (2): 665-682.

[31] Kulikova M V, Kulikov. Adaptive ODE solvers in extended Kalman filtering algorithms [J]. Journal of Computational and Applied Mathematics, 2014, 262, 205-216.

[32] Babuvška I, Rheinboldt. Error estimates for adaptive finite element computations [J]. SIAM Journal on Numerical Analysis, 1978, 15 (4): 736-754.

[33] Oden J T, Vemaganti. Estimation of local modeling error and goal-oriented adaptive modeling of heterogeneous materials: I. Error estimates and adaptive algorithms [J]. Journal of Computational Physics, 2000, 164 (1): 22-47.

[34] Vemaganti K S, Oden. Estimation of local modeling error and goal-oriented adaptive modeling of heterogeneous materials: II. a computational environment for adaptive modeling of heterogeneous elastic solids [J]. Computer Methods in Applied Mechanics and Engineering, 2001, 190 (46): 6089-6124.

[35] Zienkiewicz O C, Zhu. The superconvergent patch recovery and a posteriori error estimates. Part 1: The recovery technique [J]. International Journal for Numerical Methods in Engineering, 1992, 33 (7): 1331-1364.

[36] Zienkiewicz O C, Zhu. The superconvergent patch recovery and a posteriori error estimates. Part 2: Error estimates and adaptivity [J]. International Journal for Numerical Methods in Engineering, 1992, 33 (7): 1365-1382.

[37] Greenberg L. A prüfer method for calculating eigenvalues of self-adjiont systems of ordinary differential equations: parts 1 and 2 [J]. Technical Report. TR91-24 University of Maryland at College Park MD. , 1991.

[38] Kurochkin S V. Indexing of eigenvalues of boundary value problems for Hamiltonian systems of ordinary differential equations [J]. Computational Mathematics and Mathematical Physics, 2014, 54 (3): 439-442.

[39] Zienkiewicz O C. The background of error estimation and adaptivity in finite element computations [J]. Computer Methods in Applied Mechanics and Engineering, 2006, 195 (46): 207-213.

[40] Wilkinson J H. The Algebraic Eigenvalue Problem [J]. Clarendon Press Oxford. , 1965.

［41］ Wilkinson J H, Reinsch C. Linear algebra, handbook for automatic computation ［J］. Springer-Verlag New York. , 1971,

［42］ Bérard P, Helffer B. Sturm's theorem on zeros of linear combinations of eigenfunctions ［J］. Expositiones Mathematicae, 2020, 38 (1): 27-50.

［43］ Thomas J M. Sturm's theorem for multiple roots ［J］. National Mathematics Magazine, 1941, 15 (8): 391-394.

［44］ Ioakimidis N I. Deciding in elasticity problems by using Sturm's theorem ［J］. Computers and structures, 1996, 58 (1): 123-131.

［45］ Wiberg N E, Bausys R, Hager, et al. Adaptive h-version eigenfrequency analysis ［J］. Computers and Structures, 1999, 71 (5): 565-84.

［46］ Wiberg N E, Bausys R, Hager, et al. Improved eigenfrequencies and eigenmodes in free vibration analysis ［J］. Computers and Structures, 1999, 73 (15): 79-89.

［47］ Arndt M, Machado R D, Scremin A, et al. An adaptive generalized finite element method applied to free vibration analysis of straight bars and trusses ［J］. Journal of Sound and Vibration, 2010, 329 (6): 659-672.

［48］ Schillinger D, Rank E. An unfitted hp-adaptive finite element method based on hierarchical B-splines for interface problems of complex geometry ［J］. Computer Methods in Applied Mechanics and Engineering, 2011, 200 (4748): 3358-3380.

［49］ Bao G, Hu G, Liu D, et al. An h-adaptive finite element solver for the calculations of the electronic structures ［J］. Journal of Computational Physics, 2012, 231 (14): 4967-4979.

Chapter 3
Adaptive finite element method for vibration of non-uniform and variable curvature beams

3. 1　Introduction

Beam components are widely used in engineering applications, such as curved beam bridges in structural engineering[1], wing spar in aerospace engineering[2], and pillar Euler-Bernoulli beam model in mining engineering[3]. The mechanical characteristics of a curved beam were first studied by Ferguson[4]. Howson *et al.* studied the out-of-plane vibration of a curved beam[5]. Considerable research has been conducted on the matrix analysis of curved beam structures[6]. The evaluation of free vibrations of a beam is the basis of forced vibration research. The analysis of the natural vibration frequency can be used to avoid structural resonance[7]. Structural deformation under seismic loads can be analysed using vibration mode superposition[8]. In addition, damage identification of damaged structures can be conducted using natural frequency and vibration mode[9, 10]. The accurate identification of the number, location and degree of cracks damage depends on high-precision frequency and vibration mode solutions[11, 12].

Researchers worldwide have conducted extensive studieson curved beam structures. For example, Hartog obtained the in-plane natural vibration frequency of a circular arch using the Rayleigh-Ritz method[13]. Wu et al. deducted the free vibration equation of a Timoshenko curved beam[14, 15]. derived the decoupled vibration equation of a curved beam from the free vibration equation of an Euler curved beam[15]. Correa et al. generalised/ extended finite element analysis of curved beam vibration[16]. In the case of variable cross-section members, analytical closed solutions can be obtained for a few members with changing rules[17-19]; it is impossible to obtain exact solutions in most cases. Therefore, the finite element methodis crucial for the analysing variable cross-section members. With the development of the finite element method, many superconvergent computation methods have been developed, such as finite element interpolation post-processing[20], the local average method[21], the global smoothing method[22], the element local smoothing method[23], and element energy projection[24]. In this study, the free frequency and vibration mode of a typical plane variable cross-section and variable curvature beam members were evaluated. The cross-section of the curved beam was symmetrical at any position above the plane, and the axis of the curved beam was in the same plane. The in-plane and out-of-plane vibrations of this type of curved beam need to be decoupled and solved separately[25]. Therefore, accurate and effective solutions of the continuous order frequency and vibration mode of a variable cross-section and variable curvature beam is necessary, which is the objective of this study.

At present, the analysis methods for plane curved beam structures primarily comprisethe finite element method[26, 27], finite difference method[28], differential quadrature method[29], equivalent geometric analysis method[30], transfer matrix method[31], pseudo-spectrum method[32], and accurate dynamic stiffness method[33]. The finite element method is an effective method toobtain an approximate solution of the free vibration frequency and vibration mode of a curved beam. It is used to analyse the free vibration of complex beam components such as different cross-sections [34], support types[35], and curve linear forms[36]. Peres et al. used ageometrically accurate beam finite element to accurately predict the spatial displacement and distortion of curved beams[37]. Parallel computation and adaptive algorithms of finite element methods have been developed by many researchers[38-44]. Various new mesh adaptive algorithms utilise high-performance computation schemes to obtain high-precision solutions that meet preset error limits. Research on variable cross-sections involves investigating methods to limit the number of initial mesh divisions and local mesh refinement to mitigate some challenging problems, such as the vector Sturm-Liouville problems and free vibration of curved beams[45], free vibration and damage detection of beams[9], hp-version adaptive finite element algorithm for eigensolutions[46], and damage and fracture in rock with multi-physical field coupling[47]. This study focuses on finite element mesh refinement for in-plane and out-of-plane vibrations of variable geometrical Timoshenko beams.

In this chapter, a computation method is proposed for the superconvergent patch recovery solution of a finite elementfor determining the in-plane and out-of-plane free vibration modes of variable cross-section and variable curvature beams. The proposed method can obtain high-precision vibration mode solutions and ensure thatthe computation results meet the error limit. The remainder of this chapter is organised as follows. In Section 3.2, the partial differential governing equations of in-plane and out-of-plane free vibration of variable geometrical Timoshenko beams are introduced. In Section 3.3, finite element discretisation is presented to solve the partial differential governing equations. In Section 3.4, the superconvergent patch recovery of the vibration modes is discussed. In Section 3.5, error estimation and mesh refinement are introduced. In Section 3.6, the finite element mesh refinement strategy and procedure are introduced. In Section 3.7, numerical examples are provided to validatethe effectiveness of the proposed method in solving the continuous order frequencies and modes of in-and out-of-planefree vibration of curved beams with different curve shapes, multiple types of boundary conditions, variable curvature, and variable section forms. The main conclusions are summarised in Section 3.8.

3.2   Partial differential governing equations for in-plane and out-of-plane free vibration of variable geometrical Timoshenko beams

Consider a plane beam witha variable cross-section and curvature, as shown in Figure 3.1. The neutral axis coordinate of the curved beam is s in the coordinate system $xoyz$, where x and y are the plane coordinates of the curved beam. The plane beam is tangential tothe x-axis, normal to the y-axis and perpendicular to the plane of the z-axis.

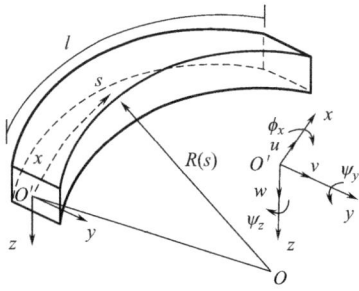

Figure 3.1 Coordinate systems and symbols of planar non-uniform and variable curvature curved beam

The displacement of the in-plane vibration is described as follows: x axial displacement amplitude is u, y axial displacement amplitude is v, and z axial rotation angle amplitude is ψ_z. The displacement of out-of-plane vibration comprises the displacement amplitude w along the z axis, the rotation angle ψ_y around the y axis, and the torsion angle φ_x around the x axis. The curvature radius of the curved beam is $R(s)$, the shear stiffness correction coefficient of the section is k, the section area is $A(s)$, the moment of inertia along the y axis is $I_y(s)$, the moment of inertia along the z axis is $I_z(s)$, the polar moment of inertia is $I_p(s)$, the torsion constant is $J(s)$, and length is l. The elastic modulus of the material is E, the shear modulus is G, Poisson's ratio is v, and the density is ρ.

In this study, the differential governing equation[48] of the in-plane free vibration of the curved plane beam is as follows:

$$-(EAu')' + (EAv/R)' + \kappa GAv'/R + \kappa GAu/R^2 - \kappa GA\psi_z/R = \omega^2 \rho Au \qquad (3.1a)$$

$$-(\kappa GAv')' - (\kappa GAu/R)' - EAu'/R + (\kappa GA\psi_z)' + EAv/R^2 = \omega^2 \rho Av \qquad (3.1b)$$

$$-(EI\psi_z')' - \kappa GAv' - \kappa GAu/R + \kappa GA\psi_z = \omega^2 \rho I_z\psi_z \qquad (3.1c)$$

The differential governing equation of the out-of-plane free vibration[33] is as follows:

$$-[\kappa GA(w' + \psi_y)]' = \omega^2 \rho Aw \qquad (3.2a)$$

$$-[EI_y(\psi_y' + \phi_x/R)]' + \kappa GA(w' + \psi_y) - (GJ/R)(\phi_x' - \psi_y/R) = \omega^2 \rho I_y\psi_y \qquad (3.2b)$$

$$-GJ(\phi_x' - \psi_y/R)' + (EI_y/R)(\psi_y' + \phi_x/R) = \omega^2 \rho I_p\phi_x \qquad (3.2c)$$

In the equations above, $(\ )' = d(\ )/ds$ and ω are the natural frequencies.

Both in-plane and out-of-plane free vibration governing equations, Equations (3.1) and (3.2), can be written as the eigenvalue equation in the form of the following matrix:

$$\boldsymbol{Lu} = \omega^2 \boldsymbol{Ru} \qquad (3.3)$$

In Equation (3.3), $\boldsymbol{u}$ is the vibration mode (displacement) function vector (in-plane and out-of-plane vibrations are $\{u, v, \psi_z\}^T$ and $\{w, \psi_y, \varphi_x\}^T$, respectively) and ω and $\boldsymbol{u}$ represent the eigenvalues and eigenvectors, respectively. $(\omega, \boldsymbol{u})$ are collectively called eigen-pairs in this study. $\boldsymbol{L}$ and $\boldsymbol{R}$ are the corresponding differential operator matrices.

3.3 Finite element discretisation

To solve the eigenvalue equation, Equation (3.3), the following eigenvalue equation of the linear matrix is established using a conventional finite element under the given finite element mesh π:

$$\boldsymbol{KD} = \omega^2 \boldsymbol{MD} \qquad (3.4)$$

In the equation above, $\boldsymbol{D}$ is the finite element solution of the modal vector and $\boldsymbol{K}$ and $\boldsymbol{M}$ are static stiffness matrices and uniform mass matrices, respectively. Usingthe inverse iteration method[49], the following finite element solution under the current mesh was obtained: $(\omega^h, \boldsymbol{u}^h)$.

3.4   Superconvergent patch recovery of vibration modes

The displacement finite element method considers the node displacement as the basic unknown quantity. The finite element displacement solution has a superconvergence of h^{m+1} order (m is the order of the current form function) inside the element and a superconvergence of the h^{2m} order on the element nodes; these nodes form the displacement superconvergent points. The error of the solution at the superconvergent points is smaller than that in other regions and has a faster convergence rate[49]. By using adjacent element segments and interpolating the displacement values of the element nodes with the $\overline{m}$ order higher-order function (higher than the current form function order m, $\overline{m} \geqslant m+1$), a superconvergent solution of the displacement field in the element domain can be obtained, and a superconvergent piecerecovery method for post-finite element processing can be established[9, 50, 51]. Under the current mesh, the displacement superconvergent solution has a higher convergence order than the conventional finite element solution. In this study, for the in-plane and out-of-plane free vibration of curved beams, after obtaining the finite element solution of vibration mode (displacement) under the current mesh, the adjacent elements of element e were pieced together using the superconvergent piece recovery method of finite element post-processing. Subsequently, the high-order function interpolation was realised. The superconvergent solution of the vibration mode can be obtained as follows:

$$u_i^*(x) = \boldsymbol{Pa}, \quad i = 1, 2, 3 \tag{3.5}$$

In the equation above, $\boldsymbol{P}$ is the given function vector, and $\boldsymbol{a}$ is the vector of undetermined coefficients.

$$\boldsymbol{P} = \begin{bmatrix} 1 & x & \cdots & x^p \end{bmatrix}, \quad \boldsymbol{a} = \begin{bmatrix} a_1 & a_2 & \cdots & a_m \end{bmatrix}^{\mathrm{T}} \tag{3.6}$$

The value of the coefficient vector $\boldsymbol{a}$ is determined by obtaining the minimum value of the following functional such that the product result in Equation (3.5) is equal to the current node displacement value at the element node:

$$\Pi = \sum_{j=1}^{n} (u_i^*(x_j) - \boldsymbol{P}(x_j)\boldsymbol{a}), \quad i = 1, 2, 3 \tag{3.7}$$

In the equation above, n is the number of nodes for assembling elements.

Using the least-squares method to solve Equation (3.8), the following value of $\boldsymbol{a}$ is obtained:

$$\boldsymbol{a} = \boldsymbol{A}^{-1}\boldsymbol{b} \tag{3.8}$$

In the equation above, each coefficient matrix is given by

$$A = \sum_{j=1}^{n} P(x_j)^{\mathrm{T}} P(x_j),$$

$$b = \sum_{j=1}^{n} P(x_j)^{\mathrm{T}} u_i^h(x_j), \quad i = 1, 2, 3 \tag{3.9}$$

The frequency value can be obtained using the mode solution and Rayleigh quotient computation[52]. The superconvergent solution of frequency can be obtained using the mode superconvergent solution ω^* obtained above:

$$\omega^* = \frac{u^{*\mathrm{T}} K u^*}{u^{*\mathrm{T}} M u^*} \tag{3.10}$$

In the equation above, $a(\ )$ and $b(\ )$ are the inner products of the strain energy and kinetic energy, respectively.

The frequency superconvergent solution computed using the Rayleigh formula has a higher convergence order than the mode superconvergent solution[52]. Therefore, in this study, error estimation and control for mode vibration were conducted to obtain a high-precision mode solution and validate the accuracy of the frequency solution.

3.5 Error estimation and mesh refinement

By introducing the superconvergent solution of the vibration mode, the error estimation of the finite element solution of the vibration mode in the form of energy mode can be conducted[9, 53]:

$$\| e^* \| \leqslant Tol \cdot [(\| u^h \|^2 + \| e^* \|^2)/n_e]^{1/2} \tag{3.11}$$

In the equation above, n_e is the number of elements for assembling and $e^* = u^* - u^h$. The energy norm of e^* is as follows:

$$\| e^* \| = [a(e^*, e^*)]^{1/2} = \left[\int_0^l e^{*\mathrm{T}} L e^* dx \right]^{1/2} \tag{3.12}$$

In Equation (3.11), the error estimation can be written in the form of the relative error as follows:

$$\xi = \frac{\| e^* \|}{\bar{e}}, \quad \bar{e} = Tol \cdot [(\| u^h \|^2 + \| e^* \|^2)/n_e]^{1/2} \tag{3.13}$$

In the equation above, ξ is the relative error value.

For the error of the vibration mode solution on the control element, ξ should satisfy

$$\xi \leqslant 1 \tag{3.14}$$

Using modal error estimation, the mesh can be optimised to reduce and control the modal error and achieve the preset solution accuracy. The error control equation evaluates each finite element modal error on the element e; if therelative error value does not satisfy the condition presented in Equation (3.14), it indicates that the error of the modal solution on the element is too large. Therefore, in this study, the element uniform subdivision refinement model was established through mesh optimisation to increase the degree of freedom and reduce the error of the solution on the element[9, 53]. The length of the new element generated by the current element subdivision is related to the current error and the order of the element; the current error can be used to estimate the length of the new element:

$$h_{\text{new}} = \xi^{-1/m} h_{\text{old}} \tag{3.15}$$

In the equation above, h_{new} is the length of the new element, and h_{old} is the length of the current cell e. Thus, the number of newly generated elements h_{new} is

$$n_{\text{new}} = h_{\text{old}} / h_{\text{new}} = \xi^{1/m} \tag{3.16}$$

To adapt to the complexity of the problem, a reliable solution can be obtained using nonuniformly distributed meshes in the solution domain. The above adaptive analysis process evenly divides and refines the elements. If too many elements are subdivided each time, it can result in redundant engineering mechanics of the final element, increasing the computation scale. In this study, the proposed method controls the number of refinements in each cell subdivision, and the actual number of uniform cell subdivisions is

$$n_{\text{new}} = \min(\lfloor \xi^{1/m} \rfloor, d) \tag{3.17}$$

In theequation above, $\lfloor \cdot \rfloor$ is the integer symbol upward, used to ensure that the vibration mode solution error is minimised in each element subdivision, and d is the limit on the number of element subdivisions. The parameter value is set by the user so that the number of control elements is not too high.

3.6   Finite element mesh refinement strategy and procedure

Based on the vibration mode of the above superconvergent solution of finite element solution error estimation andmesh subdivision refinement methods, an effective adaptive mesh analysis method can be established to evaluate the in-plane and out-of-plane vibration of variable geometrical Timoshenko beams. Subsequently, an optimised mesh can be obtained to meet the high-precision frequency and vibration model of the error limit. For the mode of vibration, the superconvergent solution $\boldsymbol{u}^*$ had a higher convergence order than the finite element solution $\boldsymbol{u}^h$ under the current mesh. Thus, $\boldsymbol{u}^*$ wascloser to the exact solution $\boldsymbol{u}$ than $\boldsymbol{u}^h$. Therefore, the error of $\boldsymbol{u}^h$ wasanalysed and controlled using $\boldsymbol{u}^*$ instead of $\boldsymbol{u}$, and the error estimation in the form of energy mode was obtained. The displacement superconvergent solution is used to calculate the Rayleigh quotient[53] to obtain the superconvergent solution of the frequency. The frequency superconvergent solution obtained using the Rayleigh equation has a higher convergence order than the vibration mode superconvergent solution. In this study, the error of the vibration mode solution was estimated and controlled to ensure that the frequency solution meets the error requirements. Thus, the shutdown criterion of the proposed algorithm was established by controlling the vibration mode error. By subdividing and refining the mesh based on the error estimation of the finite element solution, the optimised mesh, high-precision frequency, and mode of vibration solutions meeting the error limits can be obtained; these form a set of finite element adaptive solutions. Thus, based on the proposed method, the in-plane and out-of-plane vibration problems of beams, such as parabolic curved beams, beams with variable cross-sections and curvatures, elliptically curved beams, and circularly curved beams, can be solved. The procedure of finite element mesh refinement for in-plane and out-of-plane vibration of variable geometrical Timoshenko beams based on the superconvergent patch recovery solutions of vibration modes is shown in Figure 3. 2.

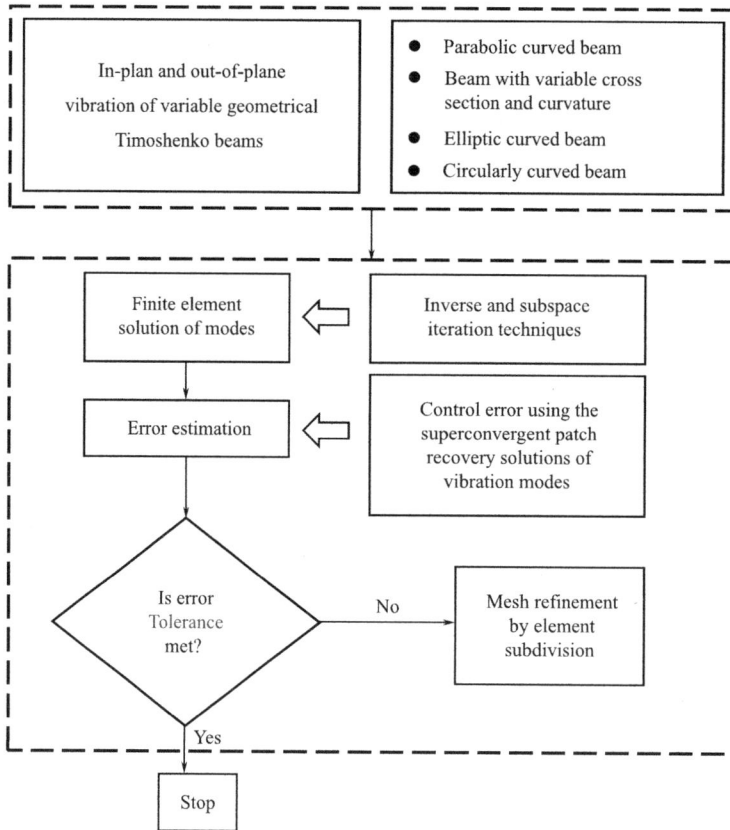

Figure 3. 2 Procedure of finite element mesh refinement for in-plane and out-of-plane vibration of variable geometrical Timoshenko beams based on the superconvergent patch recovery solutions of vibration modes

The main adaptive process includes the following steps:

(1) **Finite element solution of modes:** For the curved beam vibration problem, conventional finite element computation is carried out on the current mesh π, and the finite element solution $(\omega^h, \boldsymbol{u}^h)$ of the eigen-pair under the current mesh is obtained.

(2) **Error estimation:** The modal superconvergent solution under the current mesh can be obtained using the superconvergent patch recovery method after finite element processing; the error estimates between the superconvergent solution and the finite element solution are obtained.

(3) **Mesh refinement by element subdivision:** For the elements that do not satisfy the error

control equation, Equation (3.14), the mesh subdivision and refinement method is used to subdivide them to obtain a new finite element mesh and return to Step 1. If all the elements meet the error limits, the mesh does not need to be subdivided, and the solving process ends.

3.7　Numerical examples

The method used in this study was programmed in Fortran 9.0. In this section, numerical examples of in-plane and out-of-plane free vibrations of various beams with variable cross-sections and curvatures are provided. The accuracy and applicability of the proposed method are validatedby comparing the free vibration results of typical parabolic curved beams, beams with variable cross-section and curvature, elliptically curved beams with the free vibration results of parabolic curved beams, and curved beams with out-of-plane free vibration. Three dimensions were adopted for the computation examples in this section, and two elements were adopted for the initial mesh. The initial error limit was $Tol = 10^{-4}$, and the limit on the number of element subdivisions was set as $d = 6$.

3.7.1　Example 1: In-plane vibration of a parabolic curved beam

A parabolic curved beam of span $L = 5$ m and height $h = 4$ m is shown in Figure 3.3.

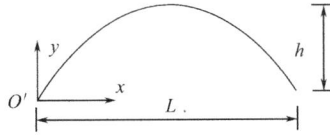

Figure 3.3　Geometric model of parabolic curved beam

The parabolic equation of the curved beam is

$$y = (-4h/L^2)x(x-L)\tag{3.18}$$

The basic parameters of the curved beam are as follows:

$$A = 0.004 \text{ m}, \ E = 70 \text{ GPa}, \ \kappa = 0.85,$$
$$\kappa G/E = 0.3, \ \rho = 2777 \text{ kg/m}^3, \ I_z = 0.0001 \text{ m}^4\tag{3.19}$$

Using the proposed method, the first three successive eigen-pairs of the curved beam were computed under different boundary conditions, such as fixed support at both ends, fixed support at one end, simple support at both ends, different height-span ratios ($h/L = 0.2$ and 0.8), and equal working conditions. The natural vibration frequencies were transformed into dimensionless values using the following equation: $\overline{\omega} = \omega L^2 \sqrt{\rho A/EI_z}$. The computed frequencies are listed in Table 3.1. The frequency values were obtained based on existing literature[54] on the above problems and utilising the dynamic stiffness method and Taylor series expansion method. The frequency values are listed in Table 3.1. By comparing the results obtained using the proposed method and the dynamic stiffness method, the results obtained using the two methods were in agreement with each other

41

under various boundary conditions. Thus, the accuracy of the proposed method and its applicability to various boundary conditions were verified. Owing to the complexity of the free vibration of curved beams with variable curvatures and cross-sections, it is difficult to obtain the exact analytical solutions of the natural vibration frequencies and modes. To verify that the solution satisfies the error limit, the first three orders of solutions, under the condition that the height-span ratio h/L of the two ends of the fixed curved beam is 0.2 and 0.8, are presented as examples. A high-precision finite element solution under a dense mesh (500 uniformly distributed elements) was used as the exact solution. The relative error between the solution of frequency of each order obtained using the proposed method and the exact solution is shown in Figure 3.4. In Figure 3.4, the maximum error $h/L = 0.8$ is the third-order frequency of working conditions, and the maximum error value is 0.21×10^{-4}; the solution of the frequency of each order meets the preset error limit of 10^{-4}.

Non-dimensional frequencies $\overline{\omega}$ of in-plane vibration of parabolic curved beam Table 3.1

h/L	k	Frequency values					
		Clamped-clamped		Clamped-hinged		Hinged-hinged	
		$\overline{\omega}_k^h$	$\overline{\omega}_k^h$ ①	$\overline{\omega}_k^h$	$\overline{\omega}_k^h$ ①	$\overline{\omega}_k^h$	$\overline{\omega}_k^h$ ①
	1	46.0245	46.0252	36.5733	36.6037	28.7587	28.7644
0.2	2	87.1265	87.2729	78.0592	78.1589	68.2433	68.3075
	3	126.287	126.280	123.343	123.3630	123.282	123.2460
	1	10.9353	10.9359	8.34175	8.34340	6.37925	6.38010
0.8	2	25.7756	25.7958	21.7083	21.7199	17.9637	17.9707
	3	45.6961	45.7441	40.3252	40.3514	35.3552	35.3749

Source: ① Results from paper [54].

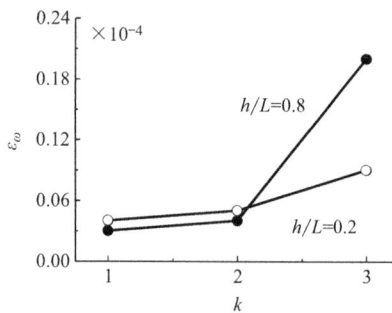

Figure 3.4　Errors of results obtained using proposed
method and high-precision solutions under dense mesh

Figures 3.5~3.7 show the results of solving the first three modes under the fixed support boundary condition and height-span ratio $h/L = 0.8$ with the proposed method. The final distribution of the adaptive mesh is marked on the horizontal axis. To facilitate intuitive display and comparative analysis, the vibration mode results in the Figure 3. are normalised (the maximum value of the vibration mode is 1); the displacement of the vibration mode at the two fixed ends is 0. In this method, non-uniform meshes were divided for all modes. Sparse meshes were used in the region with low mode variation, and relatively dense meshes were used in the region with severe mode variation to avoid the redundancy of uniform fine meshes in the whole region.

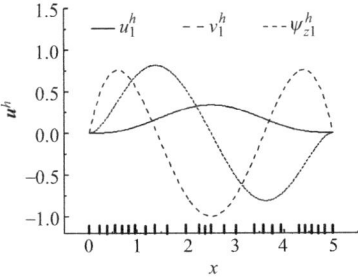

Figure 3.5   First-order vibration mode (u_1^h, v_1^h, ψ_{z1}^h)

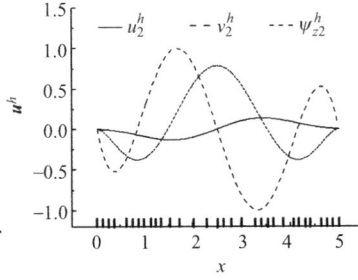

Figure 3.6   Second-order vibration mode (u_2^h, v_2^h, ψ_{z2}^h)

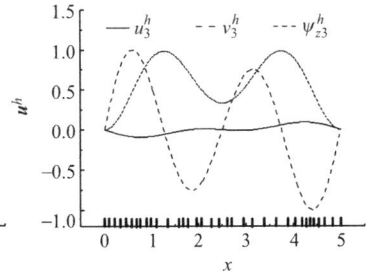

Figure 3.7   Third-order vibration mode (u_3^h, v_3^h, ψ_{z3}^h)

3.7.2   Example 2: In-plane vibration of a beam with variable cross-section and curvature

A beam with variable cross-section and curvature is shown in Figure 3.8. Both ends of the curved beam are fixed support boundary conditions, and the span is L. The beam width (along the direction perpendicular to the paper) is $b = 2\mathrm{m}$, and the upper boundary function of the curved beam is

Figure 3.8   Geometric model of non-uniform curved beam with variable curvature

$$y = 2 \qquad (3.20)$$

The lower boundary function is

$$y = -0.005x^2 + 0.2x - 2(\mathrm{m}) \qquad (3.21)$$

The neutral axis function is

$$y = -0.0025x^2 + 0.1x\,(\mathrm{m}) \qquad (3.22)$$

The basic parameters of the curved beam are

$$L = 40 \text{ m}, \ E = 70 \text{ GPa}, \ G = 24.50 \text{ GPa},$$
$$\kappa = 0.8438, \ \rho = 2777 \text{ kg/m}^3 \qquad (3.23)$$

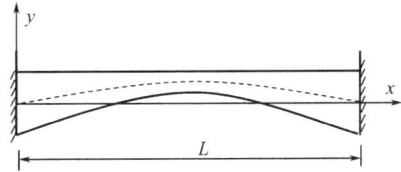

The proposed method was used to continuously solve for the first four order eigenpairs of the in-plane free vibration of the curved beam, and the computed frequency values are listed in Table 3.2. In a study by Zhao[55], the dynamic stiffness method based on the Wittrick-Williams algorithm was developed to solve this problem; the solutions are presented in Table 3.2. The results obtained using the proposed method are in good agreement with the results obtained using the dynamic stiffness method, validating the effectiveness of the proposed method for evaluating the in-plane free vibration of beams with variable cross-sections and curvature forms.

Frequencies ω of in-plane vibration of non-uniform curved beam with variable curvature

Table 3. 2

k	Frequency values	
	ω_k^h (Hz)	$\omega_k^{h\,①}$ (Hz)
1	72. 1035	72. 05
2	150. 842	150. 78
3	267. 482	267. 34
4	407. 912	407. 77

Source: ① Results from paper [55].

3. 7. 3   Example 3: In-plane vibration of an elliptically curved beam

To further test the applicability of the proposed method, the in-plane free vibration of

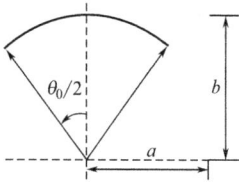

Figure 3. 9   Geometric model of elliptically curved beam

an elliptically curved beam with fixed ends is considered, as shown in Figure 3. 9. The length of the long axis and the short axis of the elliptic arc of the curved beam are $2a=5$ m and $2b=2.5$ m, respectively. The initial angle is $\theta_0=120°$, and the other parameters are the same as in Equation (3.19). The proposed method was used to solve for the first six order eigenpairs; the natural vibration frequencies were converted

into dimensionless values using $\overline{\omega}=\omega\,(2a)^2\sqrt{\rho A/EI_z}$. The results are listed in Table 3. 3. The computed results were compared with the solutions obtained using the dynamic stiffness method and Taylor series expansion method in literature[54]. The fourth-and fifth-order frequency values of the two methods were slightly different (the relative errors were 0. 0041 and 0. 0052, respectively), and the other orders had a high degree of coincidence.

Non-dimensional frequencies $\overline{\omega}$ of in-plane vibration of an elliptically curved beam

Table 3. 3

k	Frequency values		k	Frequency values	
	$\overline{\omega}_k^h$	$\overline{\omega}_k^{h\,①}$		$\overline{\omega}_k^h$	$\overline{\omega}_k^{h\,①}$
1	67. 4352	67. 5534	4	211. 537	212. 414
2	84. 0931	84. 2185	5	325. 811	327. 516
3	160. 403	160. 647	6	358. 439	358. 960

Source: ① Results from paper [54].

3. 7. 4   Example 4: Out-of-plane vibration of a parabolic curved beam

Considering the out-of-plane free vibration of the parabolic curved beam shown in Figure 3. 3 of Example 1, the parabolic equation is the same as Equation (3.18), and the basic parameters of the curved beam are as follows:

$$L = 28.87 \text{ m}, \ h = 5.774 \text{ m}, \ E = 26 \text{ GPa},$$
$$E/G = 2.6, \ \kappa = 5/6, \ \rho = 2166.7 \text{ kg/m}^3, \tag{3.24}$$
$$A = 3 \text{ m}^2, \ I_y = 0.25 \text{ m}^4, \ I_p = 2.5 \text{ m}^4$$

The proposed method was used to calculate the first seven consecutive eigen-pairs of the out-of-plane free vibration of the curved beam under different boundary conditions, such as fixed supports at both ends, fixed supports at one end, and simple supports at both ends; the results of natural vibration frequency were transformed into dimensionless values using $\overline{\omega} = \omega L^2 \sqrt{\rho A / EI_y}$. The results are listed in Table 3.4. The dynamic stiffness method based on the Wittrick-Williams algorithm[56], the Runge-Kutta method, and the Regula-Falsi method[57] was used to evaluate the out-of-plane free vibration of a beam with constant cross-section and variable curvature. The computed frequency values under different boundary conditions are listed in Table 3.4. The results obtained using the proposed method are in good agreement with those obtained using the dynamic stiffness method, the Runge-Kutta method, and the Regula-Falsi method, validating the effectiveness of the proposed method for evaluating the external free vibration of a beam with variable curvature. Note that some results (underlined) in the solutions presented in the literature differ in the ones digit, caused by the absence of strict error control in the iterative solution approximation process.

Non-dimensional frequencies $\overline{\omega}$ of out-of-plane vibration of parabolic curved beam

Table 3.4

k	Frequency values								
	Clamped-clamped			Clamped-hinged			Hinged-hinged		
	$\overline{\omega}_k^h$	$\overline{\omega}_k^h$ ①	$\overline{\omega}_k^h$ ②	$\overline{\omega}_k^h$	$\overline{\omega}_k^h$ ①	$\overline{\omega}_k^h$ ②	$\overline{\omega}_k^h$	$\overline{\omega}_k^h$ ①	$\overline{\omega}_k^h$ ②
1	17.0442	17.03	17.12	11.1276	11.12	11.15	6.08274	6.079	6.090
2	48.3965	48.37	48.77	38.9657	38.94	39.10	30.4037	30.38	30.40
3	95.0213	94.97	96.06	82.1935	82.14	82.61	70.0338	69.99	70.03
4	109.942	109.9	109.9	109.828	109.8	109.8	109.839	109.7	109.8
5	156.526	156.4	158.7	140.486	140.4	141.4	125.037	125.0	125.0
6	203.793	203.7	203.8	203.788	203.7	203.8	193.982	193.8	194.0
7	230.935	230.8	234.7	212.172	212.1	213.8	203.764	203.7	203.8

Source: ① Results from paper [56], ② Results from paper [57].

Figure 3.10 shows the seventh-order vibration mode results obtained using the proposed method under the condition that one end is fixed and one end is simply supported. The final distribution of the adaptive mesh is marked on the horizontal axis; the displacement of the fixed left end is 0, w_7^h and ψ_{y7}^h at the right simply supported end are both zero, and φ_{x7}^h is non-zero. Considering the

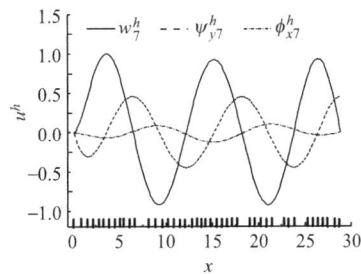

Figure 3.10   Seventh-order vibration mode (w_7^h, ψ_{y7}^h, ϕ_{x7}^h)

variation of high-order modes, the proposed method could still effectively divide non-uniform meshes. To test the reliability of mesh optimisation, relatively dense meshes were used in the areas where the modes changed significantly.

3.7.5 Example 5: Out-of-plane vibration of a circularly curved beam

A curved arc beam with fixed supports at both ends is shown in Figure 3.11. The arc angle is θ_0, and the cross-section of the beam is round. The basic parameters of the curved beam are

$$R=10 \text{ m}, \ \theta_0=60°, \ E=2.6\times10^7 \text{ Pa},$$

$$E/G=2.6, \ \kappa=0.89, \ \rho=2600 \text{ kg/m}^3, \quad (3.25)$$

$$A=\pi \text{ m}^2, \ I_y=\pi/4 \text{ m}^4, \ I_p=\pi/2 \text{ m}^4$$

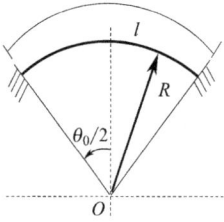

Figure 3.11 Geometric model of circularly curved beam

The proposed method was used to calculate the first four consecutive eigen-pairs of the out-of-plane free vibration of the curved beam; the natural vibration frequency was transformed into a dimensionless value using $\overline{\omega}=\omega R^2 \sqrt{\rho A/EI_y}$. The computed frequency values are presented in Table 3.5. In the literature[29, 5], the dynamic stiffness method and Bresse-Timoshenko beam theory were used to evaluate the out-of-plane free vibration of curved beams. The computational results are listed in Table 3.5. The results obtained using the proposed method are in good agreement with those presented in the literature. Thus, the accuracy of the proposed method for evaluating the external free vibration of arc-shaped curved beams was validated.

Non-dimensional frequencies $\overline{\omega}$ of out-of-plane vibration of circularly curved beam

Table 3.5

k	Frequencies values		
	$\overline{\omega}_k^h$	$\overline{\omega}_k^{h①}$	$\overline{\omega}_k^{h②}$
1	16.88516	16.88	16.885
2	39.70043	39.70	39.700
3	40.93463	40.90	40.934
4	70.58075	70.51	70.581

Source: ① Results from paper [29]; ② Results from paper [5].

3.8 Conclusions

The main conclusions of this study and the potential of the proposed computation method can be summarised as follows:

(1) Finite element mesh refinement based on the superconvergent patch recovery solutions of vibration modes was proposed for the in-plane and out-of-plane vibrations of

variable geometrical Timoshenko beams. The superconvergent solution of each order of vibration mode wasestablished; the mesh was adaptively refined based on the mode superconvergent solution. Thus, the optimised mesh and the frequency and vibration modes satisfying the error limit given by the user were obtained.

(2) The results computed using the proposed algorithm were in good agreement with those computed using other high-precision algorithms (the dynamic stiffness method, Taylor series expansion method, Runge-Kutta method, Regula-Falsi method, and Bresse-Timoshenko beam theory), validating the accuracy of the proposed algorithm for beam analysis. Through the numerical analysis of parabolic curved beams, beams with variable cross-sections and curvatures, elliptically curved beams, and circularly curved beams, the solutions of frequencies were verified to beconsistent with the results obtained using other specially developed methods, validating the applicability of the algorithm.

(3) When analysing the change in high-order vibration mode, the proposed method could perform the mesh adaptive analysis of a curved beam structure efficiently. The parts where the vibration mode changed significantly were locally densified, and a relatively fine mesh division was adopted, highlighting the reliability of mesh optimisation processing of the proposed algorithm.

The proposed algorithm can be universally applied for beam analysis by efficiently analysing the vibration problems of curved beams with different curve shapes, various boundary conditions, variable cross-sections, and variable curvatures. The algorithm can be extended for application in superconvergent computation and adaptive analysis of finite element solutions of general structures and solid deformation fields; in addition, it can be used for adaptive analysis of more complex plates, shells, and three-dimensional structures. Furthermore, this method can be potentially used toanalyse the vibration and stability of curved members with crack damage, obtain high-precision vibration modes and instability modes under damage defects, and provide reliable solutions for analysing the influence of crack damage on curved members; relevant research progress will be reported in the future.

References

[1] Seo J, Linzell D G. Horizontally curved steel bridge seismic vulnerability assessment [J]. Engineering Structures, 2012, 34: 21-32.

[2] Guo S, Li D, Wu J, et al. Theoretical and experimental study of a piezoelectric flapping wing rotor for micro aerial vehicle [J]. Aerospace Science and Technology, 2012, 23 (1): 429-438.

[3] Please C P, Mason D P, Khalique C M, et al. Fracturing of an Euler-Bernoulli beam in coal mine pillar extraction [J]. International Journal of Rock Mechanics and Mining Sciences, 2013, 64: 132-138.

[4] Kapania R K, Li J. A formulation and implementation of geometrically exact curved beam elements incorporating finite strain and finite rotations [J]. Computation Mechanics, 2003, 30 (5/6): 444-459.

[5] Howson W P, Jemah A K. Exact out-of-plane natural frequencies of curved Timoshenko beams [J]. Journal of Engineering Mechanics, 1999, 125 (1): 19-25.

[6] Zhang L. Structural matrix analysis of curved beam [J]. Journal of The China Railway Society, 1993, 23: 80-86.

[7] Hu W H, Thöns S, Rohrmann R G, et al. Vibration-based structural health monitoring of a wind turbine system. Part I: Resonance phenomenon [J]. Engineering Structures, 2015, 89: 260-272.

[8] De Domenico D, Falsone G, Ricciardi G, et al. Improved response-spectrum analysis of base-isolated buildings: A substructure-based response spectrum method [J]. Engineering Structures, 2018, 162: 198-212.

[9] Wang Y, Ju Y, Zhuang Z, et al. Adaptive finite element analysis for damage detection of non-uniform Euler-Bernoulli beams with multiple cracks based on natural frequencies [J]. Engineering Computations, 2018, 35 (3): 1203-1229.

[10] Chang K C, Kim C W. Modal-parameter identification and vibration-based damage detection of a damaged steel truss bridge [J]. Engineering Structures, 2016, 122: 156-173.

[11] Fan W, Qiao P. Vibration-based damage identification methods: A review and comparative study [J]. Structural Health Monitoring, 2011, 10 (1): 83-111.

[12] Yang Y B, Yang J P. State-of-the-art review on modal identification and damage detection of bridges by moving test vehicles [J]. International Journal of Structural Stability and Dynamics, 2018, 18 (2): 850025.

[13] Hartog J P D X L. The lowest natural frequency of circular arcs [J]. Philosophical Magazine, 1960, 5 (28): 400-408.

[14] Wu J S, Lin F T, Shaw H J, et al. Free in-plane vibration analysis of a curved beam (arch) with arbitrary various concentrated elements [J]. Applied Mathematical Madelling, 2013, 37 (14/15): 7588-7610.

[15] Li X, Li P, Sun L, et al. Decoupling solution and verification of differential equation of free vibration of curved beam [J]. Nuclear Power Engineering, 2016, 37 (S2): 7-10.

[16] Correa R M, Arndt M, Machado R D, et al. Free in-plane vibration analysis of curved beams by the generalized/extended finite element method [J]. European Journal of Mechanics, 2021, 88: 104244.

[17] Rossi R E, Laura P A A, Gutierrez R H, et al. A note on transverse vibration of Timoshenko beam of non-uniform thickness clamped at one end and carrying a concentrated mass at the other [J]. Journal of Sound and Vibration, 1992, 143: 491-502.

[18] Lee S Y, Ke H Y, Kuo Y H, et al. Analysis of non-uniform beam vibration [J]. Journal of Sound and Vibration, 1990, 142: 15-29.

[19] Lau J H. Vibration frequencies for a non-uniform beam with end mass [J]. Journal of Sound and Vibration, 1984, 97 (3): 513-521.

[20] Mclean P, Leger P, Tinawi R, et al. Post-processing of finite element stress fields using dual kriging based methods for structural analysis of concrete dams [J]. Finite Elements in Analysis and Design, 2006, 42 (6): 532-546.

[21] Bramble J H, Schatz A H. Higher order local accuracy by averaging in the finite element method [J]. Mathematical Aspects of Finite Elements in Partial Differential Equations, 1974, 31 (137): 1-14.

[22] Oden J T, Demkowicz L, Rachowicz W, et al. Toward a universal h-p adaptive finite element strategy, Part 2. A posteriori error estimation [J]. Computer Methods in Applied Mechanics and

Engineering, 1989, 77 (1): 113-180.

[23] Bdoroomand B, Zienkiewicz O C. Recovery by equilibrium in patches (REP) [J]. International Journal for Numerical Methods in Engineering, 1997, 40: 137-64.

[24] Yuan S, Ye K, Wang Y, et al. Adaptive finite element method for eigensolutions of regular second and fourth order Sturm-Liouville problems via the element energy projection technique [J]. Engineering Computations, 2017, 34 (8): 2862-2876.

[25] Chidamparam P, Leissa A W. Vibrations of planar curved beams, rings, and arches [J]. Applied Mechanics Reviews, 1993, 46 (9): 467-483.

[26] Yang F, Sedaghati R, Esmailzadeh E, et al. A finite thin circular beam element for out-of-plane vibration analysis of curved beams [J]. Journal of Mechanical Science and Technology, 2009, 23 (5): 1396-1405.

[27] Krishnan A, Suresh Y J. A simple cubic linear element for static and free vibration analyses of curved beams [J]. Computers and Structures, 1998, 68 (5): 473-489.

[28] Wankui B, Hui X. Curved beam elasticity theory based on the displacement function method using a finite difference scheme [J]. Springer Nature Journal, 2019, 2019: 1-8.

[29] Kang K, Bert C W, Striz A G, et al. Vibration analysis of shear deformable circular arches by the differential quadrature method [J]. Journal of Sound and Vibration, 1995, 183 (2): 353-360.

[30] Tao-An H, Anh-Tuan L, Jaehong L, et al. Bending, buckling and free vibration analyses of functionally graded curved beams with variable curvatures using isogeometric approach [J]. Sejong University, 2017, 52: 2527-2546.

[31] Bickford W B, Storm B T. Vibration of plane curved beams [J]. Journal of Sound and Vibration, 1975, 39 (2): 135-146.

[32] Jinhee L. Free vibration analysis of circularly curved multi-span Timoshenko beams by the pseudospectral method [J]. Journal of Mechanical Science and Technology, 2007, 21 (12): 2066-2072.

[33] Huang C S, Tseng Y P, Chang S H, et al. Out-of-plane dynamic analysis of beams with arbitrarily varying curvature and cross-section by dynamic stiffness matrix method [J]. International Journal of Solids and Structures, 2000, 37 (3): 495-513.

[34] Alshorbagy A E, Eltaher M A, Mahmoud F F, et al. Free vibration characteristics of a functionally graded beam by finite element method [J]. Applied Mathematical Modelling, 2011, 35 (1): 412-425.

[35] Villavicencio R, Soares C G. Numerical modelling of the boundary conditions on beams stuck transversely by amass [J]. International Journal of Impact Engineering, 2011, 38 (5): 384-396.

[36] Piovan M T, Domini S, Ramirez J M, et al. In-plane and out-of-plane dynamics and buckling of functionally graded circular curved beams [J]. Composite Structures, 2012, 94 (11): 3194-3206.

[37] Peres N, Goncalves R, Camotim D, et al. A geometrically exact beam finite element for curved thin-walled bars with deformable cross-section [J]. Computer Methods in Applied Mechanics and Engineering, 2021, 381.

[38] Arndt M, Machado R D, Scremin A, et al. An adaptive generalized finite element method applied to free vibration analysis of straight bars and trusses [J]. Journal of Sound and Vibration, 2010, 329 (6): 659-672.

[39] Arndt M, Machado R D, Scremin A, et al. Accurate assessment of natural frequencies for uniform and non-uniform Euler-Bernoulli beams and frames by adaptive generalized finite element method [J]. Engineering Computations, 2016, 33 (5): 1586-1609.

[40] Zhang L, Zheng W, Lu B, et al. Parallel adaptive finite element software platform PHG and its

application［J］. Science China Press, 2016, 46 (10): 1442-1464.

［41］ Zhang L. A parallel algorithm for adaptive local refinement of tetrahedral meshes using bisection［J］. Global-Science Press, 2009, 2 (1): 65-89.

［42］ Liu H, Zhang L. Common marking strategy of adaptive finite element and its parallel implementation in PHG［J］. Journal On Numerical Methods and Computer Applications, 2009, 30 (4): 315-320.

［43］ Cheng J, Zhang L. Parallel adaptive finite element software PHG solid for three-dimensional structural analysis［J］. Computer Science, 2012, 39 (5): 278-281.

［44］ Liu H, Leng W, Cui T, et al. Research on load balancing in parallel adaptive finite element calculation ［J］. Journal On Numerical Methods and Computer Applications, 2015, 36 (3): 166-184.

［45］ Wang Y. An *h*-version adaptive FEM for eigenproblems in system of second order ODEs: Vector Sturm-Liouville problems and free vibration of curved beam［J］. Engineering Computations, 2020, 37 (1): 1210-1225.

［46］ Wang Y, Wang J. An *hp*-version adaptive finite element algorithm for eigensolutions of free vibration of moderately thick circular cylindrical shells via error homogenization and higher-order interpolation ［J］. Engineering Computations DOI: 101108/EC-07-2021-0430, 2021.

［47］ Wang Y. Adaptive analysis of damage and fracture in rock with multiphysical fields coupling［M］. Springer Press, 2021.

［48］ Friedman Z, Kosmatka J B. An accurate two-node finite element for shear deformable curved beams ［J］. International Journal for Numerical Methods in Engineering, 1998, 41: 473-498.

［49］ Zienkiewicz O C, Taylor R L, Zhu J Z, et al. The Finite Element Method: Its Basis and Fundamentals, 7th Ed. ,［M］. Oxford: Elsevier (Singapore) Pte Ltd, 2015.

［50］ Wiberg N E, Bausys R, Hager P, et al. Adaptive *h*-version eigenfrequency analysis［J］. Computers and Structures, 1999, 71 (5): 565-84.

［51］ Wiberg N E, Bausys R, Hager P, et al. Improved eigenfrequencies and eigenmodes in free vibration analysis［J］. Computers and Structures, 1999, 73 (1/5): 79-89.

［52］ Clough R W, Penzien J. Dynamics of Structures［M］. 2nd Edtion. New York: McGraw-Hill, 1993.

［53］ Zienkiewicz O C, Zhu J. The superconvergent patch recovery and a posteriori error estimates. Part 2: Error estimates and adaptivity［J］. International Journal for Numerical Methods in Engineering, 1992, 33 (7): 1365-1382.

［54］ Tseng Y P, Huang C S, Lin C J, et al. Dynamic stiffness analysis for in-plane vibrations of arches with variable curvature［J］. Journal of Sound and Vibration, 1997, 207: 15-31.

［55］ Zhao X. Research on dynamic stiffness method for free vibration of planar curved beams［D］. Beijing: Tsinghua University, 2010.

［56］ Ye K and Zhao X. Dynamic stiffness method for out-of-plane free vibration analysis of planar curved beam［J］. Engineering Mechanics, 2012, 29 (3): 1-8.

［57］ Lee B K, Oh S J, Mo, et al. Out-of-plane free vibrations of curved beams with variable curvature［J］. Journal of Sound and Vibration, 2008, 318 (1/2): 227-246.

Chapter 4
Adaptive finite element method for vibration disturbance of cracked beams

4. 1　Introduction

As a fundamental component, circularly curved beams are widely used in civil engineering, mechanical engineering, aerospace engineering, and other related fields[1, 2]. In practical engineering, the curved beams often undergo crack damage. Therefore, the accurate evaluation of the dynamic performance of curved beams with cracks is a critical consideration in structural design. More specifically, the depth, number, and distribution of crack damage would change the fundamental characteristics of curved beams, disturbing the frequency and vibration mode of the beams[3-5], Consequently, it affects the reliability of the structure, and the stress concentration at the crack tip will cause the crack to continue expanding, eventually leading to structural failure. Therefore, studying the dynamic characteristics of a structure with cracks, particularly the free vibration characteristics, can effectively ensure the safe use and targeted reinforcement of the structure by helping to accurately identify, maintain, and replace cracks before they reach the critical value. Moreover, it can avoid huge potential safety hazards to a great extent. Simultaneously, the actual natural frequency and mode of vibration of a damaged curved beam can be used to identify and locate cracks[6, 7]. The accurate identification of crack depth, number, and location depends on the high-precision solution of frequency and mode shape[8, 9], which has become a challenging problem. Numerous analytical methods and theoretical models have been developed to obtain high-precision free vibration solutions for beam members. However, it is still difficult to effectively apply to complex conditions, such as variable curve line-version, multi-crack damage, and various boundary conditions[10, 11].

Since the initial microcracks in the early structure are not easily identified, they are easily ignored. However, the in-depth propagation of microcracks often leads to major catastrophic accidents. In a structural element, cracks in the form of initial defects within the beam or caused by fatigue or stress concentration reduce the natural frequencies and change the vibration mode shapes owing to the local flexibility introduced by the crack. The fracture of bridges, the collapse of mines, and the fracture and leakage of oil and gas pipelines incur huge losses. Therefore, monitoring and predicting the existence[12-14] and depth of early cracks[15-21] and preventing major accidents are critical research directions in the field of damage identification. Kisa[22] adopted the finite element method to study the effects of the location and depth of cracks on the natural frequency and vibration mode of transverse non-propagating open crack beams by using the drop in the natural frequencies and the change in the mode shapes. Accordingly, the presence and nature of cracks in a structure can be detected. Similarly, the research progress of crack identification methods based on modal analysis and intelligent methods is introduced. On this basis, existing problems and research development directions for structural crack identification are proposed. Rizos *et al.* [17] proposed a method to determine the crack location and depth of the beam by using the bending vibration mode of the cantilever beam. Gounaris *et al.* [23] established a method for identifying the crack location and depth in a beam based on the vibration mode difference between the cracked beam and the intact beam. Rahimi et al. [24] studied the free vibration of functionally graded beams with transverse cracks under clamped-clamped boundary conditions based on the Reddy high-order theory. The authors also examined the effects of the slender ratio, location of the crack, depth of the crack, and material gradients on natural frequency.

Most of the aforementioned studies have focused on single-crack beams. The number and distribution of cracks also have a great influence on the vibration characteristics of beams[25-30]. Hu and Liang[31] proposed a technique for detecting multiple cracks. The continuum damage model was adopted to identify the discretising elements of a structure that contained cracks. Subsequently, the spring damage model was employed to quantify the location and size of the discrete crack in each damaged element. Moreover, Zheng and Fan[32] computed the natural frequencies of a Timoshenko beam with an arbitrary number of transverse open cracks. Lee[33] presented a simple, straightforward method to identify multiple cracks in a beam. The method is based on the massless rotational spring model for cracks, the finite element method, and the Newton-Raphson method. The approximate crack location was obtained by Armon's rank-ordering method[34], using the first four eigenfrequencies, and the crack size was determined using the finite element method. Shin *et al.* [35] examined the natural vibrations of a beam resting on an elastic foundation with a finite number of transverse open cracks. Frequency equations were derived for beams with different end restraints. The effects of the crack location, size, number, and foundation constants on the natural frequencies of the beam were investigated. Sawant *et al.* [36] adopted the finite element model of a cantilever beam before and after damage, set cracks with different depths at different locations, and investigated the association between the change in the natural frequency, distribution, and depth of the crack.

Numerical computation has increasingly become a reasonable choice and a vital technology in the analysis of the dynamic performance of complex structures. Li[37] employed the asymptotic analysis of the Timoshenko load theory to examine the free vibration of an axial shear beam bearing a concentrated mass at the end of an elastic support by considering the rotational motion of the section. Sina *et al.* [38] utilised a beam theory different from the traditional first-order shear deformation beam theory to analyse the free vibration of functionally graded beams. Kou and Yang[39] applied the meshless boundary domain integral equation method to analyse the free vibration of functionally graded beams with edge cracks. Labib *et al.* [40] incorporated the Wittrick-Williams algorithm and the dynamic stiffness method to examine the free vibration. However, their analytical method is only applicable to beams with uniform cross-sections. The finite element method has been developed and applied to solve the natural frequency and vibration mode of curved beams with cracks[41-43]. Kisa *et al.* [44] integrated the finite element method and component mode synthesis to analyse the free vibration characteristics of cracked Timoshenko beams. Chinchalkar[45] adopted the approximation of a spring with rotational stiffness within the conventional finite element method to solve the free vibration problem for cracked wedges and two-segment beams. However, the accuracy of the solution depends on the quality of meshing. Therefore, inevitably, the solution introduces errors due to meshing[46]. The conventional finite element method cannot satisfy the requirements of computation accuracy and mesh convenience. In particular, the disturbance effect of crack damage on the vibration modes of each order demands higher requirements for the effectiveness of nonuniformly distributed meshes. The finite element mesh adaptive analysis method can effectively optimise the mesh distribution and obtain non-uniform meshes. The process reduces the finite element computation cost and significantly improves the computation accuracy by seeking the optimal solution in line with the accuracy. The adaptive finite element method exhibits good solution effects for a series of challenging problems, namely elastic buckling of non-uniform Euler-Bernoulli beams[47], vector Sturm-Liouville problems, free vibration of curved beams[48], damage detection of beams[49], hp-version adaptive finite element algorithm for eigensolutions[50], and damage and fracture in rock with multi-physical fields coupling[51].

In this study, the damage defect analogy scheme of a circularly curved beam was established, where the crack depth, number, and location, were simulated. The h-version finite element mesh adaptive analysis method[48] of Timoshenko beam with variable cross-section is introduced to solve the free vibration problem of a circularly curved beam with crack damage. The optimised mesh and high-precision continuous order natural frequency and mode of vibration of the preset error tolerance can be obtained. In this analysis, a numerical example for solving the free vibration of a circularly curved beam with multi-crack damage is presented. The disturbance effects of multi-crack damage depth, number, and distribution on the natural frequency and vibration mode of circularly curved beams were analysed using the obtained frequency difference and vibration mode difference with or without damage.

The remainder of this study proceeds as follows. Section 4.2 introduces the characterisation method of crack damage of circularly curved beams. In Section 4.3, the free vibration equation of a circularly curved beam is introduced, and the development of the finite element solution by using the inverse power iteration method is described. Section 4.4 introduces the error estimation technology in the adaptive algorithm program and the mesh adaptive refinement encryption method. In Section 4.5, numerical examples are presented to demonstrate the solution of the free vibration of a variety of typical circularly curved beams with cracks to test the reliability and practicability of the proposed algorithm. Lastly, a summary is provided of the main conclusions of numerical simulation.

54

4.2   Characterisation method for microcrack damage in circularly curved beams

When consideringthe plane curved beam with crack damage shown in Figure 4.1, the neutral axis coordinates of the curved beam are s and the coordinate system is $xoyz$, where x and y are in-plane coordinates of the curved beam, x is tangential along the axis, y is normal along the axis, and z is perpendicular to the plane where the axis lies. The displacement of in-plane vibration is as follows: displacement amplitude u along the x axis, displacement amplitude v along the y axis, and the rotational amplitude ψ_z around the z axis. The curvature radius of the curved beam is $R(s)$, the section shear stiffness correction coefficient is κ, the section area is denoted by $A(s)$, the inertia moment

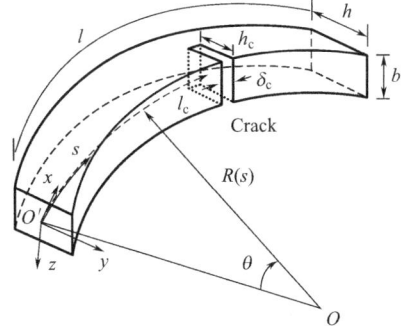

Figure 4.1   Coordinate systems and symbols of a cracked curved beam

of the z axis is $I(s)$, the length is l, beam height is h, and the beam thickness is b. The elastic modulus of the material is denoted as E, the shear modulus is G, the Poisson's ratio is ν, and the density is ρ.

In this study, themicrocrack damage in a curved beam was studied, which weakened the section of the beam and attenuated the section properties of the beam. In this study, the crack section damage defect analogy method[49] and the crack section damage is defined as

$$\alpha = \frac{h_c}{h} \tag{4.1}$$

$$\beta = \frac{l_c}{l} \tag{4.2}$$

where α denotes the section damage rate that characterises the crack damage depth (size), and $\alpha=0$refers to the beam section being intact, β represents the section location of the crack damage, and $h_c$and l_c denote the absolute depth of the crack and the coordinate value of the centre axis, respectively. The cross-sectional inertia moment and area of the beam section are weakened based on the damage rate of the crack section.

$$I_c = \frac{bh^3(1-\alpha)^3}{12} \tag{4.3}$$

$$A_c = bh(1-\alpha) \tag{4.4}$$

where I_c and A_c denote the moment of inertia and the area of the cross-section considering crack damage, respectively. The section damage width δ_c of microcracks is very small (or does not consider the crack width). Moreover, δ_c was set as follows to control the crack width without affecting the accuracy of the adaptive analysis results

$$\delta_c = 0.01 \times Tol \tag{4.5}$$

where Tol represent the preset error tolerance of free vibration solution.

4.3　Partial differential governing equations and finite element discretisation for free vibration of circularly curved beams

In this study, the in-plane free vibration of a curved plane beam was studied. The differential governing equations of the eigenvalue problem are as follows[48,·52] :

$$-(EA\boldsymbol{u}')' + (EAv/R)' + \kappa GAv'/R + \kappa GA\boldsymbol{u}/R^2 - \kappa GA\psi_z/R = \omega^2 \rho A\boldsymbol{u}$$
$$-(\kappa GAv')' - (\kappa GA\boldsymbol{u}/R)' - EA\boldsymbol{u}'/R + (\kappa GA\psi_z)' + EAv/R^2 = \omega^2 \rho Av \tag{4.6}$$
$$-(EI\psi_z')' - \kappa GAv' - \kappa GA\boldsymbol{u}/R + \kappa GA\psi_z = \omega^2 \rho I\psi_z$$

where $()' = d\ ()/ds$, ω denotes the natural frequency; $\boldsymbol{u} = \{u, v, \psi_z\}^{\mathrm{T}}$ is a vector corresponding to the vibration mode(displacement)function, ω and $\boldsymbol{u}$ correspond to the eigenvalues and eigenvectors, respectively. In this study, $(\omega, \boldsymbol{u})$ is collectively called an eigenpair.

The free vibration control equation (Equation (4.6)) can be written as the eigenvalue equation in the form of a matrix as follows:

$$\boldsymbol{Lu} = \omega^2 \boldsymbol{Ru} \tag{4.7}$$

where $\boldsymbol{L}$ and $\boldsymbol{R}$ are the corresponding differential operator matrices.

To solvethe eigenvalue problem represented by Equation (4.7), under the given finite element mesh π, the conventional finite element establishes the following linear matrix eigenvalue equation:

$$\boldsymbol{KD} = \omega^2 \boldsymbol{MD} \tag{4.8}$$

where $\boldsymbol{D}$ represents the finite element solution of the vibration mode vector, $\boldsymbol{K}$ and $\boldsymbol{M}$ denote the static stiffness matrix and uniform mass matrix, respectively. The finite element solution $(\omega^h, \boldsymbol{u}^h)$ under the current mesh is obtained using the inverse power iteration method[53] .

4.4　Local mesh refinement techniques and procedure

In the finite element computation, thereare superconvergent points[53] with a higher convergence order compared with the current mesh solution. The accuracy of the current finite element solution can be improved by using superconvergent points combined with element patching and high-order shape function interpolation technology, and a global superconvergent solution[49, 54, 55] can be obtained. In this study, the finite element solution of the vibration mode (displacement) under the current mesh is obtained for the free vibration problem of a circularly curved beam, and the superconvergent solution of the vibration mode is obtained using the finite element post-processing superconvergent patch recovery method.

$$w^*(x) = Pa \tag{4.9}$$

where P represents the given function vector, and a denotes the undetermined coefficient vector. Subsequently, the natural frequency can be obtained using the vibration mode solution and Rayleigh quotient computation[56] . The error estimation in the form of energy mode can be performed for the finite element solution of the vibration mode under the current mesh by introducing the superconvergent solution of the vibration mode[49, 53] .

$$\xi = \frac{\|e^*\|}{\bar{e}} < 1, \ \bar{e} = Tol \cdot [(\|w^h\|^2 + \|e^*\|^2)/n_e]^{1/2} \tag{4.10}$$

where ξ denotes the relative error value, n_e 、 corresponds to the number of elements for patching, e^* 、 $= w^* - w^h$ and $\|e^*\|$ is the energy norm.

The use of vibration mode error estimation enables optimisation of the mesh to reduce and control the vibration mode error and achieve the preset solution accuracy. In this study, the vibration mode error of each finite element e was determined. If the error control expressed by Equation (4.10) is not satisfied, the error of the vibration mode solution on the element is too large. Therefore, it was necessary to optimise the mesh. In this analysis, the h-version mesh adaptive method of uniform subdivision and encryption of the element are adopted to increase the freedom of the model and reduce the error of the solution on the element[49] . The new element length generated by the current mesh subdivision is related to the current error and element order. In other words, the present error can be used to estimate the length of the new mesh.

$$h_{new} = \xi^{-1/m} h_{old} \tag{4.11}$$

where h_{old} is the length of the current element e , and h_{new} is the length of the new element.

Based on the above methods, the following overallcomputation and analysis scheme can be formed to obtain high-precision solutions of each order frequency and vibration mode of a circularly curved beam with crack damage, and the free vibration disturbance and local mesh refinement induced by microcrack damage can be detected. The analysis procedure is illustrated in Figure 4.2 and implemented via the following the steps:

(1) **Circularly curved beam model with crack damage:** The depth, number, and distribution of multiple cracks in a curved beam are simulated using the crack damage characterisation method (Equations (4.1)~(4.5)) to form a circularly curved beam model with crack damage.

(2) **Finite element solution of frequency and vibration mode under the current mesh:** In the current finite element mesh, the finite element inverse power iterative analysis method (Equations (4.6)~(4.8)) for the free vibration problem of a curved beam is used to solve the circularly curved beam model with crack damage and obtain the finite element solution of frequency and vibration mode.

(3) **Error estimation and updating the finite element mesh:** The mesh adaptive

subdivision encryption method is used to estimate the error of the current vibration mode solution. When the preset error tolerance remain unsatisfied, mesh subdivision encryption is performed in the vibration mode region of the crack damage disturbance to obtain the updated encryption mesh (Equations (4.9) $\sim$ (4.11)). Under the updated finite element mesh, return to steps (2) and (3) for cyclic computation and error estimation until a set of fully optimised meshes and solutions satisfying the error tolerance are obtained.

(4) **Effects of crack damage on free vibration disturbance:** The size, location, and number of different crack damages are analysed, and the disturbance effects of damage and conditions on the free vibration frequency and vibration mode are studied.

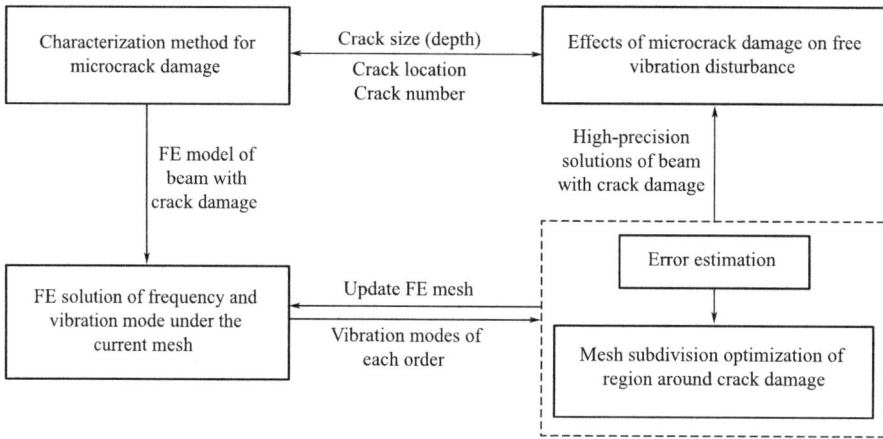

Figure 4.2   Analysis procedure of free vibration disturbance and local mesh refinement induced by microcrack damage in circularly curved beams

4.5   Numerical examples

The method in thisstudy compiles the corresponding Fortran 90 language program code. This section presents several representative free vibration numerical examples of circularly curved beams with cracks. Moreover, the accuracy of the mesh adaptive division, frequency, and vibration solution are discussed. The influence of crack damage depth, number, distribution, and other factors on the natural frequency and vibration mode disturbance were analysed to verify the reliability and practicability of the algorithm. In this section, all the examples use three elements, and the initial mesh uses two elements. The initial error limit given is $Tol=10^{-4}$.

4.5.1   Example 1: Verification for eigensolutions of free vibration of uncracked curved beam

The constant section 1/4 circularly curved beams with fixed ends shown in Figure 4.3 is solved to test the accuracy and effectiveness of the proposed method in solving uncracked circularly curved beam. The geometric and physical parameters of the curved beam are as follows:

$$R=1, \ EI=1/12, \ EA=\rho=\kappa GA=1 \quad (4.12)$$

Table 4.1 lists the first 5 order vibration frequency solutions obtained using the method proposed herein. The high-precision solutions obtained using nine thousand linear elements with constant cross-sections by an analytical method[57] were compared and analysed. Simultaneously, the number of final elements

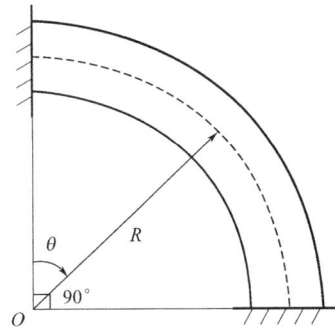

Figure 4.3　Model of a quarter of uncracked circularly curved beam

and frequency errors for each order is given. It is evident that the solutions obtained by the proposed method under the adaptive mesh are far less than the preset error tolerance requirements. It should be pointed out that the error control of the vibration mode was performed in this study, and the frequency with higher convergence order is obtained by using the vibration mode solution and Rayleigh quotient[56] to ensure that the frequency value can also strictly satisfy the error tolerance.

Frequencies of a quarter of uncracked circularly curved beam　　　Table 4.1

k	Frequency (Hz)		Number of elements	Error $(\times 10^{-6})$
	ω_k^h	$\omega_k^{h \ ①}$		
1	1.64129401	1.64129400	20	0.004
2	1.89492153	1.89492042	20	0.383
3	2.88881791	2.88881787	28	0.010
4	3.92909842	3.92909842	32	0.205
5	4.46223401	4.46223400	32	0.002

Source: ① Results from paper [57].

Figure 4.4 illustrates the first-and fifth-order modal solutions obtained using the proposed method. The final distribution of the adaptive mesh is marked on the transverse axis. The vibration mode results in Figure 4.4 were normalised (the maximum vibration mode value was set to 1) to facilitate intuitive display and comparative analysis. It is evident that the displacement of vibration mode at both fixed ends is 0; In the present analysis, non-uniform meshes are divided for each order of vibration modes, and sparse meshes are used in the region where the vibration mode changes gently, and relatively fine meshes are used in the region where the vibration mode changes violently, which avoids the redundancy of using uniform fine meshes in the entire region. The complexity of the vibration mode increases with an increase in the order, and the fifth-order vibration mode uses more elements than the first-order vibration mode.

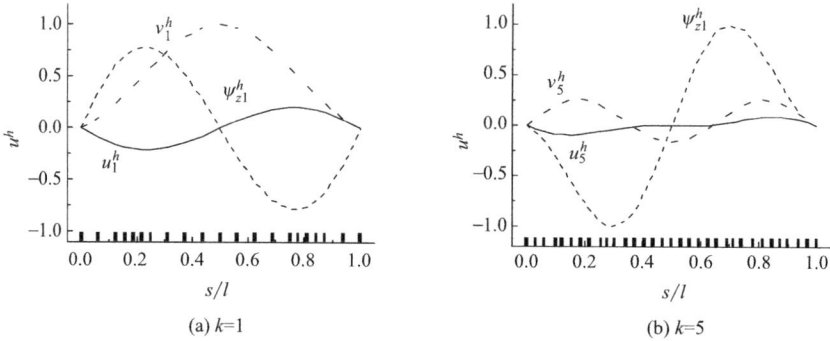

Figure 4. 4 Vibration modes of a quarter of uncracked circularly curved beam

4. 5. 2 Example 2: Different depths of single-crack beams

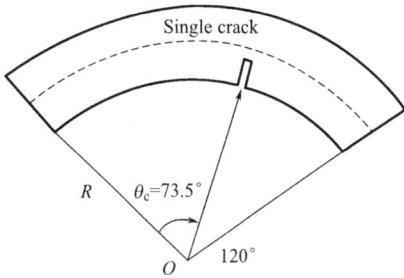

Figure 4. 5 Model of a curved beam with a single crack in different depth cases

$(\alpha = 0.0, 0.16, 0.5; \beta = 0.6125)$

As shown in Figure 4. 5, the present study solves the single-crack damage curved beams with simply supported ends to test the effectiveness of the proposed method to analyse the free vibration of circularly curved beams with different crack depths. The crack depths of the curved beam are $\alpha = 0.0$, 0.16, 0.5, and the crack locations are $\beta = 0.6125$. The angle of the circularly curved beam was $120°$, and the crack location angle was $73.5°$. The other fundamental geometric and physical parameters are as follows:

$$R = 1.0 \text{ m, } b = 0.015 \text{ m, } b = 0.045 \text{ m, } E = 206 \text{ GPa, } \rho = 7850 \text{ kg/m}^3 \quad (4.13)$$

The proposed method was used to calculate the continuous first 5 order eigenpairs of the free vibration of the curved beam under three crack damage depths. The computed frequency values are presented in Table 4. 2. Karaagac et al. [41] analysed the above problems by combining the energy method and the finite element model, and Cerri et al. [5] used a physical model to analyse the above line problems, the frequency values are listed in Table 4. 2. The comparison of the results from the proposed method and the energy method show that with the increase in the complexity of the problem (such as the increase in the order and the increase in the crack depth), the error of the individual order solution (marked by the underscore) increases slightly, primarily owing to the difference in the basic analysis strategy between the two methods, and it exhibits good consistency in the other orders of all types of crack depth conditions.

Frequencies of curved beam with a single crack in different depth cases　　Table 4. 2

Crack depth	Frequency	k				
		1	2	3	4	5
$\alpha=0.0$	$\omega^h_{0.0,\,k}$	24.44506	61.69788	119.07079	188.20437	277.18047
	$\omega^h_{0.0,\,k}{}^{\textcircled{1}}$	24.45	61.73	119.19	188.51	277.87
	$\omega^h_{0.0,\,k}{}^{\textcircled{2}}$	24.52	61.78	118.05	184.11	269.02
	Error $(\times10^{-2})$	0.02	0.05	0.10	0.16	0.25
$\alpha=0.16$	$\omega^h_{0.0,\,k}$	24.42725	61.67788	118.8333	188.2017	276.7535
	$\omega^h_{0.0,\,k}{}^{\textcircled{1}}$	24.44	61.72	119.16	188.49	277.85
	$\omega^h_{0.0,\,k}{}^{\textcircled{2}}$	24.48	61.72	117.76	184.03	268.24
	Error $(\times10^{-2})$	0.05	0.07	0.27	0.15	0.39
$\alpha=0.5$	$\omega^h_{0.0,\,k}$	24.21468	61.35527	116.4156	188.1066	272.5737
	$\omega^h_{0.0,\,k}{}^{\textcircled{1}}$	24.34	61.68	118.27	188.09	277.25
	$\omega^h_{0.0,\,k}{}^{\textcircled{2}}$	24.32	61.6	116.43	183.81	267.94
	Error $(\times10^{-2})$	0.49	0.52	<u>1.55</u>	0.88	<u>1.68</u>

Source: ①Results from paper [41], ②results from paper [5].

Figure 4. 6 illustrates the results of the influence of the crack depth on the natural frequency disturbance using the proposed method. The frequency value (left longitudinal axis marking scale) and frequency difference (right longitudinal axis marking scale, the difference between the frequency value with crack damage and that without crack damage) are shown in Figure 4. 6 to facilitate intuitive display and comparative analysis. It can be observed that the frequency value increases with the order, and there is no significant difference. Moreover, the frequency difference is negative, indicating that the occurrence of crack damage reduces the frequency value of each order. When $\alpha=0.5$, the decrease in the amplitude of frequency difference of each order is larger than that of $\alpha=0.16$. Besides, it can also be seen that a larger crack damage depth translates to a weaker beam section. Also, a greater attenuation degree of beam section property is manifested as a significant reduction of the frequency value.

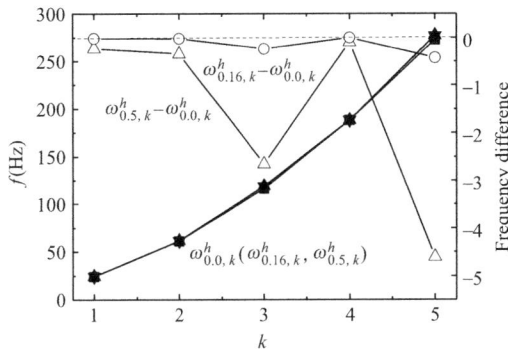

Figure 4. 6　Influence of crack depth on frequencies

Figure 4.7 illustrates that the first-order and fifth-order vibration mode solutions of single-crack damage depth $\alpha=0.5$ are obtained using the proposed method. It is evident that the vibration mode is disturbed in the vicinity of the crack damage, and the crack damage is the most obvious disturbance to the rotational displacement ψ_z. The proposed method adaptively divides the final non-uniform mesh, and a relatively dense mesh is adopted in the vicinity of the crack to adapt to the change of the vibration mode caused by the crack damage, which reflects the adaptability of the proposed method to the change of the vibration mode.

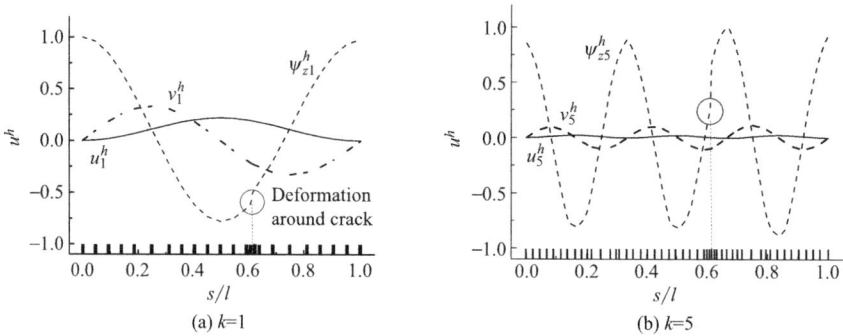

Figure 4.7　Vibration modes of curved beam with single crack in depth case $\alpha=0.5$

Figure 4.8 depicts the difference curve between each vibration mode and the corresponding non-destructive vibration mode to facilitate the analysis of the disturbance behaviour of the vibration mode caused by the degree of crack damage when the crack damages $\alpha=0.16$. It is clear that the local area of the crack damage has a significant impact on the change in the vibration mode, and the crack damage is the primary factor affecting the disturbance change of each vibration mode component. In this example, the crack damage disturbs the vibration mode components, and the rotational displacement ψ_z exhibits the largest disturbance.

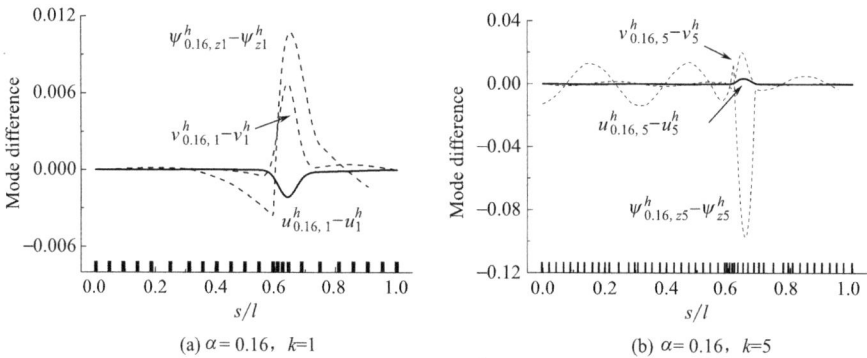

Figure 4.8　Vibration modes disturbance of curved beam with single crack in depth case $\alpha=0.16$

Figure 4.9 presents the difference curve between the vibration mode and the corresponding order of the undamaged vibration mode when the crack damage is $\alpha=0.5$.

In contrast with the vibration mode difference when $\alpha=0.16$, the vibration mode is larger. It can be inferred that the greater the degree of crack damage, the more intense is the vibration mode disturbance. The aforedescribed results indicate that the amplitude of the vibration mode difference is related to the degree of damage, and the vibration mode disturbance can be effectively controlled by quantitatively controlling the amount of crack damage. The illustrated example verifies the accuracy of the proposed method in solving the problem of a circularly curved beam with crack damage and its applicability to various problems related to crack damage depth.

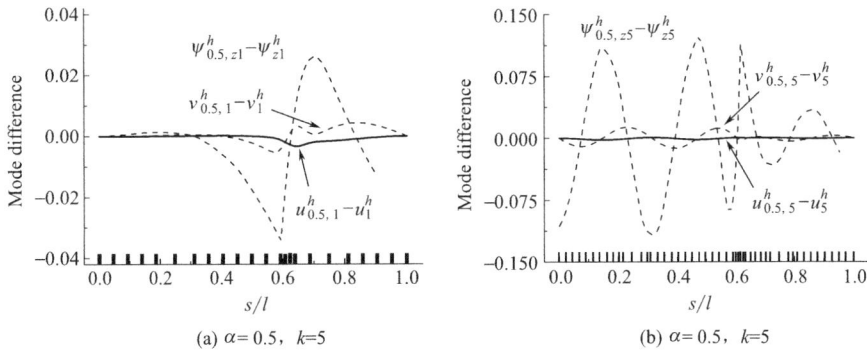

(a) $\alpha=0.5,\ k=5$ (b) $\alpha=0.5,\ k=5$

Figure 4.9 Vibration modes disturbance of curved beam with single crack in depth case $\alpha=0.5$

4.5.3 Example 3: Different numbers of multiple-crack damage

The crack number n of curved beams with crack damage is another crucial factor affecting the vibration characteristics of cracks. In this example, the typical multi-crack number ($n = 2, 3, 4$) and crack location conditions were set as shown in Table 4.3. The illustrated example uses simply supported circularly curved beams at both ends. The geometric model and basic physical parameters of the curved beams are identical to those in Equation (4.13).

Number and location of multiple-crack damage   **Table 4.3**

Cases	Crack number n	Crack distribution β
I	2	0.3333, 0.6667
II	3	0.2500, 0.5000, 0.7500
III	4	0.2000, 0.4000, 0.6000, 0.8000

The multiple cracks in each working condition have a uniform distribution form, as shown in Figure 4.10, and the angles between the cracks are $40°$ (working condition I, $n = 2$), $30°$ (working condition II, $n = 3$), and $24°$ (working condition III, $n = 4$).

The methodintroduced in the present analysis is used to compute the continuous first 50 order eigenpairs corresponding to the free vibration of the curved beam under three types of crack damage numbers. The selected typical computation frequency values are

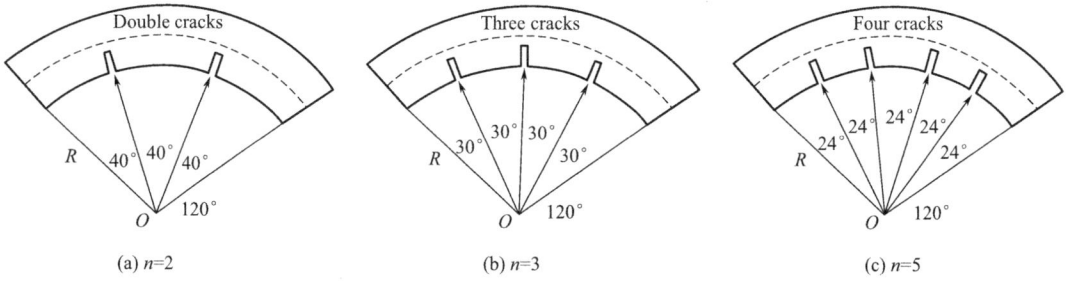

(a) $n=2$ (b) $n=3$ (c) $n=5$

Figure 4.10　Model of curved beam with multiple cracks in different number cases ($n=2$, 3, 4)

listed in Table 4.4. It is evident that an increase in the number of cracks is accompanied by a slight increase in the frequency value, except for the individual order (highlighted by the underline), and the overall trend is gradually reduced.

Frequencies of curved beam with multiple cracks in different number cases　　Table 4.4

k	Frequency (Hz)		
	$n=2$	$n=3$	$n=4$
1	48.53533213	47.90207758	47.34302395
2	153.6607502	146.6512751	144.7289066
3	299.1209131	310.0411950	292.3541860
4	489.6498928	482.8735070	505.8249710
5	751.5396271	719.6740705	712.0419733
10	2545.1390090	2515.6686390	2502.8702830
20	7706.3633080	7603.1767850	7520.4095130
30	16025.9432000	16158.5522500	16094.7618100
40	26133.8889100	26108.5497200	26050.5982100
50	38202.4438600	37858.8243400	37728.9351500

　　Figure 4.11 illustrates the results of the influence of the number of cracks on the first 5 order natural frequency disturbance. The illustration shows that the frequency value increases with the order, and there is no significant difference in the number of cracks. An increase in the number of cracks reduces the frequency value of each order. Generally, the greater the number of cracks, the greater is the degree of reduction. In the third and fourth order, the frequency difference is positive, that is, the number of cracks increases but the frequency value increases. In the fourth order, the frequency difference under the condition of crack number 4 is smaller than that under the condition of crack number 3. In other words, the increase in the crack does not significantly reduce the frequency

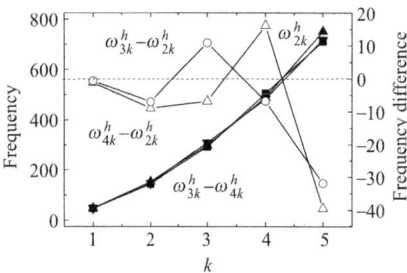

Figure 4.11　Influence of crack
number on frequencies

value. Based on the above results, it is clear that the number and location of multiple cracks affect the frequency value simultaneously. Overall, increasing the number of cracks tends to reduce the frequency of each order. However, because of the change in crack location, the frequency value would increase in some order. Therefore, based on the original multi-crack damage location, it is necessary to continue increasing the new crack or increase the original crack depth (i. e. , example 2 in this study), so that the frequency values of each order decrease.

Figure 4. 12 illustrates the first-order mode perturbation solution of a multi-crack damaged curved beam with different crack numbers solved using the proposed method. It is evident that the vibration mode is disturbed near the multi-crack damage, and the rotation displacement disturbance is the most significant. The proposed solution adaptively divides the final non-uniform mesh, and a relatively dense mesh is used near the crack to adapt to the change in vibration mode caused by crack damage, which verifies the adaptability of the adaptive meshing method to changes in vibration modes of multi-crack damaged curved beams.

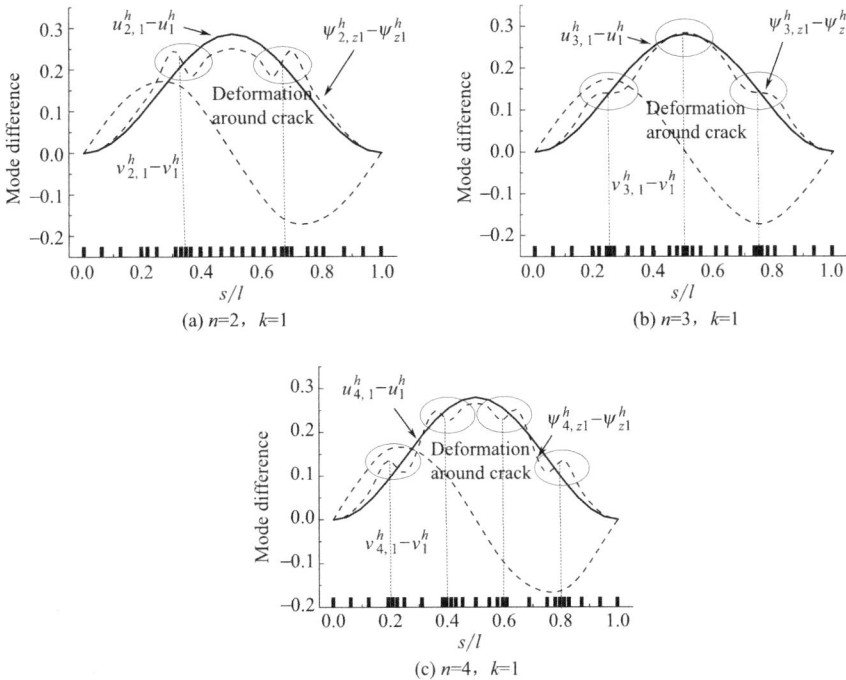

Figure 4. 12　Vibration modes disturbance of curved beam with multiple cracks in different number cases

4. 5. 4　Example 4: Different distributions of multiple-crack damage

A circularly curved beam with simply supported ends is used in this example to further analyse the influence of multi-crack damage distribution on the free vibration of a curved beam. The geometric model and fundamental physical parameters of the curved beam are

65

identical to those in Equation (4.13). The curved beam was set up with five cracks, considering the uniform distribution of crack damage along the curved beam (each crack location corresponds to β of 0.1667, 0.3333, 0.5000, 0.6667, and 0.8333), and the concentrated distribution of crack damage on the left side of the curved beam (each crack location corresponds toβ of 0.0833, 0.1667, 0.2500, 0.3333, and 0.4167). The multiple cracks in each condition are shown in Figure 4.13, and the angles between the cracks are 20° (uniform distribution of cracks) and 10° (concentrated distribution of cracks on the left side).

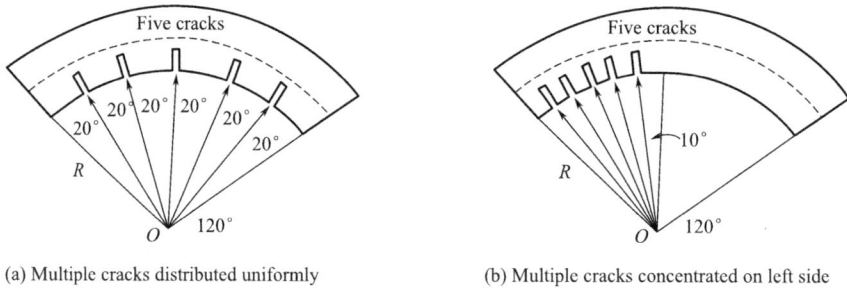

| (a) Multiple cracks distributed uniformly | (b) Multiple cracks concentrated on left side |

Figure 4.13　Model of curved beam with multiple cracks in different distribution cases

The proposed method was used to compute the continuous first 50 order eigenpairs of the free vibration of the curved beam under different crack distribution conditions. The typical computed frequency values selected are listed in Table 4.5. The influence of the crack distribution on the natural frequency disturbance is illustrated in Figure 4.14. It can be seen that the concentrated distribution on the left side of the crack exhibits more uniformity than that on the left side. Moreover, the frequency value has a higher value in the low order (such as the first to fourth orders) and has a lower value in the high order. The different distribution forms of the same number of multiple cracks have become a critical factor affecting the vibration characteristics. Therefore, it is necessary to simultaneously detect the number of cracks accurately and the location of each crack to accurately estimate the frequency value of the crack damage.

Frequencies of curved beam with multiple cracks in different distribution cases Table 4.5

k	Frequency		
	ω_{uk}^h, Multiple cracks distributed uniformly	ω_{ck}^h, Multiple cracks concentrated on left side	$\omega_{ck}^h - \omega_{uk}^h$
1	46.80879967	47.10685643	0.29806
2	143.0367853	144.70491070	1.66813
3	288.8690772	290.34379010	1.47471
4	471.6432689	477.60184400	5.95858
5	751.7922260	703.90290580	−47.88932
10	2461.5366510	2513.56899100	52.03234

continued

k	Frequency		
	ω_{uk}^{h}, Multiple cracks distributed uniformly	ω_{ck}^{h}, Multiple cracks concentrated on left side	$\omega_{ck}^{h} - \omega_{uk}^{h}$
20	7578. 2765950	7536. 88767300	−41. 38892
30	16051. 3215300	16049. 90019000	−1. 42134
40	26044. 7208000	25984. 65345000	−60. 06735
50	37833. 7406000	37557. 16953000	−276. 57107

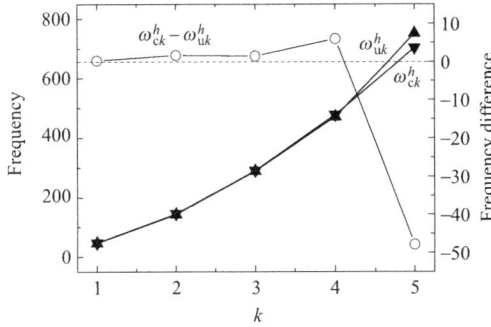

Figure 4. 14　Influence of crack distribution on frequencies

Figure 4. 15 displays the first-order vibration mode perturbation solution of a curved beam with multi-crack damage under different numbers of cracks. It is clear that the vibration mode is disturbed near the crack damage with a uniform distribution and a concentrated distribution. In this study, a relatively dense mesh is used to adapt to the change in vibration mode caused by crack damage, which reflects that the adaptive meshing method can adapt well to the vibration mode disturbance induced by different crack densities. It should be pointed out that under the condition of a concentrated distribution on the left side of the crack shown in Figure 4. 15 (b), the vibration mode at the right side exhibits a larger vibration mode disturbance concurrently, which reflects the strong disturbance behaviour of multi-crack damage to the overall vibration mode.

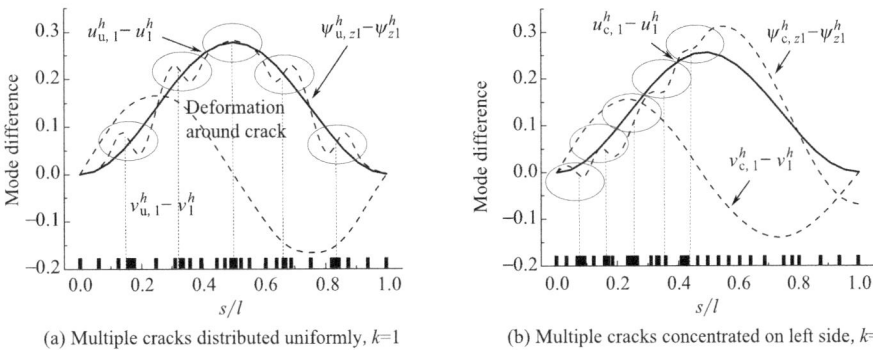

(a) Multiple cracks distributed uniformly, $k=1$

(b) Multiple cracks concentrated on left side, $k=1$

Figure 4. 15　Vibration modes disturbance of curved beam with multiple cracks in different distribution cases

4. 6   Conclusions

In this study, a cross-section damage defect comparison scheme ofa circularly curved beam with cracks and the h-version finite element mesh adaptive analysis method was established to solve the free vibration problem of a circularly curved beam with cracks. The optimised mesh and high-precision natural frequency and vibration mode solutions that satisfy the preset error tolerance were obtained for the reliable computation of the locations and sizes of multiple cracks. The disturbance behaviour of multi-crack damage depth, number, and distribution on the natural frequency and vibration mode of a circularly curved beam was quantitatively studied. The main conclusions drawn from this study are as follows:

(1) The adaptive mesh algorithm has good applicability in the analysis of uncracked and cracked curved beams and can obtain solutions that satisfy the error tolerance. The adaptive method divides the non-uniform mesh for each vibration mode, which avoids the redundancy of using a uniform fine mesh in the entire region. Curved beams with cracks were analysed, the vibration mode was disturbed near the crack damage, and the frequency value decreased. A relatively dense mesh was used near the crack to adapt to the change in vibration mode caused by crack damage, which demonstrates the accuracy and effectiveness of the proposed method.

(2) Crack damage reduces the frequency value of each order. The greater the crack damage depth, the weaker is the beam section and the greater is the attenuation of the beam section properties; this indicates that the frequency value decreases significantly, and the crack damage disturbs each vibration mode component, among which the rotational displacement ψ_z disturbance is the largest. Moreover, the greater the degree of damage, the more intense is the vibration mode disturbance.

(3) The number and location of cracks simultaneously affect the frequency value. Generally, increasing the number of cracks tends to reduce the frequency. However, as the change in crack location would also increase the frequency value of some order frequencies, the vibration modes near each crack damage are disturbed, and the disturbance in the rotational displacement ψ_z is the most significant.

(4) The concentrated distribution on one side of the crack exhibited more uniformity than that on the other side. Moreover, the frequency value has a higher value at a low order and a lower value at a high order. The vibration mode was disturbed in the vicinity of cracks with uniform and concentrated distributions, and the different distribution forms of the same number of multiple cracks become a critical factor affecting the vibration characteristics.

The proposed combination of methodologies presents a robust approach for free vibration and damage detection of beams with cracks. The proposed method reduces the cost of computation and improves the accuracy of the solutions for determining the locations and sizes of cracks in beams. The non-uniform mesh refinement in the adaptive method has high adaptability to the change in vibration mode induced by crack damage, and a high-precision vibration mode and frequency can be obtained. The present study is limited to Timoshenko beams with cracks; the findings can be used to solve for the natural frequency of plates, shells, three-dimensional structures, and heterogeneous rock mass in future studies.

References

[1] Oviedo-Tolentino F, Pérez-Gutiérrez F, Romero-Méndez R, et al. Vortex-induced vibration of a bottom fixed flexible circular beam [J]. Ocean Engineering, 88 (463-471), 2014.

[2] Fam A, Rizkalla S. Large scale testing and analysis of hybrid concrete/composite tubes for circular beam-column applications [J]. Construction and Building Materials, 2003, 17 (6-7): 507-516.

[3] Yang J, Chen Y, Xiang Y, et al. Free and forced vibration of cracked inhomogeneous beams under an axial force and a moving load [J]. Journal of Sound and Vibration, 2008, 312 (1/2): 166-181.

[4] Piovan M T, Domini S, Ramirez J M I, et al. In-plane and out-of-plane dynamics and buckling of functionally graded circular curved beams [J]. Composite Structures, 2012, 94 (11): 3194-3206.

[5] Cerri M N, Dilena M, Ruta G C, et al. Vibration and damage detection in undamaged and cracked circular arches: Experimental and analytical results [J]. Journal of Sound and Vibration, 2008, 314 (1/2): 83-94.

[6] Lee J W. Crack identification method for tapered cantilever pipe-type beam using natural frequencies [J]. International Journal of Steel Structures, 2016, 16 (2): 467-476.

[7] Rizos P F, Aspragathos N, Dimarogonas A D, et al. Identification of crack location and magnitude in a cantilever beam from the vibration modes [J]. Journal of Sound and Vibration, 1990, 138 (3): 381-388.

[8] Fan W, Qiao P. Vibration-based damage identification methods: a review and comparative study [J]. Structural health monitoring, 2011, 10 (1): 83-111.

[9] Yang Y, Yang J P. State-of-the-art review on modal identification and damage detection of bridges by moving test vehicles [J]. International Journal of Structural Stability and Dynamics, 2018, 18 (02): 1850025.

[10] Duan Y C, Wang J P, Wang J Q, et al. Theoretical and experimental study on the transverse vibration properties of an axially moving nested cantilever beam [J]. Journal of Sound and Vibration, 2014, 333 (13): 2885-2897.

[11] Jaworski J W, Dowell E H. Free vibration of a cantilevered beam with multiple steps: Comparison of several theoretical methods with experiment [J]. Journal of Sound and Vibration, 2008, 312 (4/5): 713-725.

[12] Lin H P. Direct and inverse methods on free vibration analysis of simply supported beams with a crack [J]. Engineering Structures, 2004, 26 (4): 427-436.

[13] Yang J, Chen Y. Free vibration and buckling analyses of functionally graded beams with edge cracks [J]. Composite Structures, 2008, 83 (1): 48-60.

[14] Zheng D Y, Kessissoglou N J. Free vibration analysis of a cracked beam by finite element method [J].

Journal of Sound and Vibration, 2004, 273 (3): 457-475.

[15] Nandwana B P, Maiti S K. Detection of the location and size of a crack in stepped cantilever beams based on measurements of natural frequencies [J]. Journal of Sound and Vibration, 1997, 203 (3): 435-446.

[16] Kisa M, Gurel M A. Free vibration analysis of uniform and stepped cracked beams with circular cross sections [J]. International Journal of Engineering Science, 2007, 45 (2/8): 364-380.

[17] Song M, Gong Y, Yang J, et al. Free vibration and buckling analyses of edge-cracked functionally graded multilayer graphene nanoplatelet-reinforced composite beams resting on an elastic foundation [J]. Journal of Sound and Vibration, 2019, 458 (89): 89-108.

[18] Akbaş Ş D. Free vibration of edge cracked functionally graded microscale beams based on the modified couple stress theory [J]. International Journal of Structural Stability and Dynamics, 2017, 17 (3): 1750033.

[19] Douka E, Hadjileontiadis L J. Time-frequency analysis of the free vibration response of a beam with a breathing crack [J]. NDT and E International, 2005, 38 (1): 3-10.

[20] Khoram-Nejad E, Moradi S, Shishesaz M, et al. Free vibration analysis of the cracked post-buckled axially functionally graded beam under compressive load [J]. Journal of Computational Applied Mechanics, 2021, 52 (2): 63-77.

[21] Wang T, Noori M, Altabey W A, et al. Identification of cracks in an Euler-Bernoulli beam using Bayesian inference and closed-form solution of vibration modes [J]. Proceedings of the Institution of Mechanical Engineers Part L: Journal of Materials: Design and Applications, 2021, 235 (2): 421-438.

[22] Kisa M. Free vibration analysis of a cantilever composite beam with multiple cracks [J]. Composites Science and Technology, 2004, 64 (9): 1391-1402.

[23] Gounaris G D, Papdopoulos C A. Analytical and experimental crack identification of beam structures in air or in fluid [J]. Computers and Structures, 1997, 65 (5): 633-639.

[24] Rahimi A, Livani M, Negahban Boron A, et al. Free vibration analysis of functionally graded material beams with transverse crack [J]. Journal of Mechanical Engineering, 2021, 51 (1): 277-281.

[25] Attar M. A transfer matrix method for free vibration analysis and crack identification of stepped beams with multiple edge cracks and different boundary conditions [J]. International Journal of Mechanical Sciences, 2012, 57 (1): 19-33.

[26] Wei D, Liu Y, Xiang Z, et al. An analytical method for free vibration analysis of functionally graded beams with edge cracks [J]. Journal of Sound and Vibration, 2012, 331 (7): 1686-1700.

[27] Rajasekaran S, Khaniki H B. Free vibration analysis of bi-directional functionally graded single/multi-cracked beams [J]. International Journal of Mechanical Sciences, 2018, 144: 341-356.

[28] Yoon H I, Son I S, Ahn S J, et al. Free vibration analysis of Euler-Bernoulli beam with double cracks [J]. Journal of mechanical science and technology, 2007, 21 (3): 476-485.

[29] Lien T V, Duc N T, Khiem N T, et al. Free vibration analysis of multiple cracked functionally graded Timoshenko beams [J]. Latin American Journal of Solids and Structures, 2017, 14: 1752-1766.

[30] Zheng D Y and Fan S C. Natural frequencies of a non-uniform beam with multiple cracks via modified Fourier series [J]. Journal of Sound and Vibration, 2001, 242 (4): 701-717.

[31] Hu J, Liang R Y. An integrated approach to detection of cracks using vibration characteristics [J]. Journal of the Franklin Institute, 1973, 330 (5): 841-853.

[32] Zheng D Y, Fan S. Natural frequency changes of a cracked Timoshenko beam by modified Fourier series [J]. Journal of Sound and Vibration, 2001, 246 (2): 297-317.

[33] Lee J. Identification of multiple cracks in a beam using natural frequencies [J]. Journal of Sound and Vibration, 2009, 320 (3): 482-490.

[34] Lee Y S, Chung M J. A study on crack detection using eigenfrequency test data [J]. Computers and Structures, 2000, 77 (3): 327-342.

[35] Shin Y, Yun J, Seong K, et al. Natural frequencies of Euler-Bernoulli beam with open cracks on elastic foundations. [J]. Journal of Mechanical Science and Technology, 2006, 20 (4): 467-472.

[36] Sawant S U, Chauhan S J, Deshmukh N N, et al. Effect of crack on natural frequency for beam type of structures [J]. American Institute of Physics Conference Series, 2017, AIP Publishing LLC.

[37] Li X F. Free vibration of axially loaded shear beams carrying elastically restrained lumped-tip masses via asymptotic timoshenko beam theory [J]. Journal of Engineering Mechanics, 2013, 139 (4): 418-428.

[38] Sina S A, Navazi H M, Haddadpour H, et al. An analytical method for free vibration analysis of functionally graded beams [J]. Materials and Design, 2009, 30 (3): 741-747.

[39] Kou K, Yang Y. A meshfree boundary-domain integral equation method for free vibration analysis of the functionally graded beams with open edged cracks [J]. Composites Part B: Engineering, 2019, 156 (1): 303-309.

[40] Labib A, Kennedy D, Featherston C, et al. Free vibration analysis of beams and frames with multiple cracks for damage detection [J]. Journal of Sound and Vibration, 2014, 333 (20): 4991-5003.

[41] Palash D, Talukdar S. Modal characteristics of cracked thin walled unsymmetrical cross-sectional steel beams curved in plan [J]. Thin-Walled Structures, 2016, 108: 75-92.

[42] Karaagac C, Ozturk H, Sabuncu M, et al. Crack effects on the in-plane static and dynamic stabilities of a curved beam with an edge crack [J]. Journal of Sound and Vibration, 2011, 330 (8): 1718-1736.

[43] Bovsunovskii O A. A finite-element model for the study of vibration of a beam with a closing crack [J]. Strength of Materials, 2008, 40 (5): 584-589.

[44] Kisa M, Brandon J, Topçu M, et al. Free vibration analysis of cracked beams by a combination of finite elements and component mode synthesis methods [J]. Computers and Structures, 1998, 67 (4): 215-223.

[45] Chinchalkar S. Determination of crack location in beams using natural frequencies [J]. Journal of Sound and Vibration, 2001, 247 (3): 417-429.

[46] Melenk J M, Wohlmuth B I. On residual-based a posteriori error estimation in hp-FEM [J]. Advances in Computational Mathematics, 2001, 15 (1): 311-331.

[47] Yuan S, Wang Y, Ye K, et al. An adaptive FEM for buckling analysis of non-uniform Bernoulli-Euler members via the element energy projection technique [J]. Mathematical Problems in Engineering, 2013, 40 (7): 221-239.

[48] Wang Y. An h-version adaptive FEM for eigenproblems in system of second order ODEs: Vector Sturm-Liouville problems and free vibration of curved beam [J]. Engineering Computations, 2020, 37 (1): 1210-1225.

[49] Wang Y, Ju Y, Zhuang Z, et al. Adaptive finite element analysis for damage detection of non-uniform Euler-Bernoulli beams with multiple cracks based on natural frequencies [J]. Engineering Computations, 2017, 35 (3): 1203-1229.

[50] Wang Y, Wang J. An hp-version adaptive finite element algorithm for eigensolutions of free vibration of moderately thick circular cylindrical shells via error homogenization and higher-order interpolation [J]. Engineering Computations, 2021.

[51] Wang Y. Adaptive analysis of damage and fracture in rock with multiphysical fields coupling [M]. Springer Press, 2021.

[52] Kosmatka J, Friedman Z. An accurate two-node finite element for shear deformable curved beams [J]. International Journal for Numerical Methods in Engineering, 1998, 41: 473-498.

[53] Zienkiewicz O C, Taylor R L, Zhu J Z, et al. The finite element method: its basis and fundamentals, 7th Ed, [M]. Oxford: Elsevier (Singapore), 2015.

[54] Wiberg N E, Bausys R, Hager P, et al. Adaptive h-version eigenfrequency analysis [J]. Computers and Structures, 1999, 71 (5): 565-584.

[55] Wiberg N E, Bausys R, Hager P, et al. Improved eigenfrequencies and eigenmodes in free vibration analysis [J]. Computers and Structures, 1999, 73 (1/5): 79-89.

[56] Clough R W, Penzien J. Dynamics of structures, 2nd ed [M]. New York: McGraw-Hill, 1993.

[57] Yuan S, Ye K, Wang K, et al. A self-adaptive fem for free vibration analysis of planar curved beams with variable cross-sections [J]. Engineering Mechanics, 2009, 26 (2): 126-132.

Chapter 5
Adaptive finite element method for damage detection of cracked beams

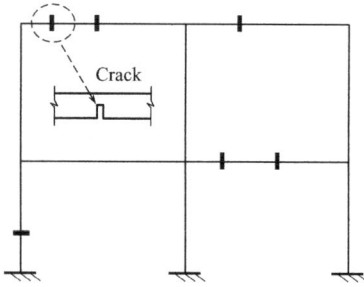

Figure 5.1 Frame structure
with multiple cracks

5.1 Introduction

Free vibration and damage detection problems for frame structures are widespread in engineering practice. In one of the special cases, the beam members contain multiple cracks, and the existence of these cracks changes the mechanical properties of the entire structure, thereby affecting its safety and applicability. Exploration of the dynamic characteristics and identification of crack locations and sizes for frame structures with multiple cracks, as shown in Figure 5.1, effectively guarantees the safety of the structures throughout their life cycles. Based on the dynamic characteristics of a cracked structure, there are well-developed damage detection methods[1, 2]. Damage detection methodologies based on actual measured frequencies are practical and effective for evaluating frame structures with multiple cracks[3-5].

As shown in Figure 5.2, the free vibration of a beam with cracks is a forward eigenproblem that involves solving for the frequencies and modes based on knowledge of the material properties. Correspondingly, damage detection is an inverse eigenproblem that involves solving for the material properties (i.e. cracks) and modes based on knowledge of the actual frequencies[6]. The present chapter addresses both forward eigenproblems and inverse eigenproblems.

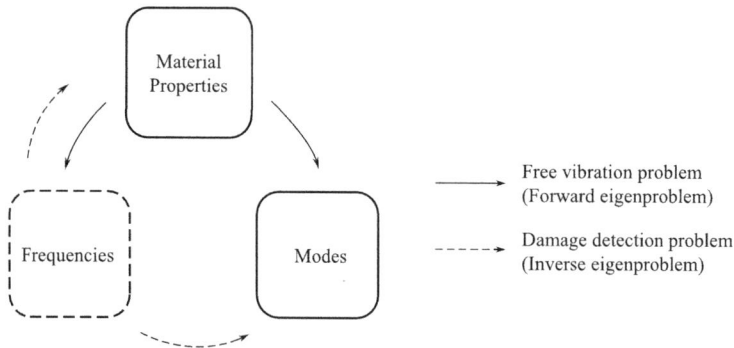

Figure 5.2 Free vibration problem (forward eigenproblem) and damage detection problem
(inverse eigenproblem) of a beam with multiple cracks

In recent years, several other methods have been proposed that are dedicated to free vibration forward eigenproblems. Methods for free vibration of uncracked beams are well developed: e. g. Euler-Bernoulli beams and shear-flexible arches with various cross-section depths and various types of supports[7-9]; consequently, the corresponding forward eigenproblems for cracked beams naturally became the next research target. Labib et al. [10] used the exact dynamic method and the Wittrick-Williams algorithm to solve the free vibration of beams and frames with multiple cracks; the method is applied effectively for uniform beams. Nandwana and Maiti[11] and Chaudhari and Maiti[12] used semi-analytical methods for solving free vibration problems for cracked beams. To the best of the authors' knowledge, SLEUTH[13] is the only code that specifically solves beam members based on Euler-Bernoulli beam theory in the challenging form of a regular fourth-order eigenproblem. SLEUTH uses piecewise constant approximations of the variable coefficients in fourth-order eigenproblems with shooting methods used to locate eigenvalues. Unfortunately, this code does not impose error control on the eigenfunctions; hence, it cannot serve as a complete eigensolver. Caddemi and Morassi[14] used Heaviside and Dirac's delta distribution functions to solve vibration problems of Euler-Bernoulli beam with multiple cracks. Caddemi and Caliò[15] proposed an exact procedure for the reconstruction of multiple instances of concentrated damage on a straight beam. Hsu[16] formulated the eigenvalue problems for clamp-free and double-hinged Euler-Bernoulli beams with elastic foundations, a single edge crack, axial loading, and excitation force by using the differential quadrature method. Rizos et al. [17] simulated cracks in beam members as springs with rotational stiffness and used the actual frequencies to identify damage; their results were in good agreement with the experimental analysis, so the crack model has widely been used in subsequent finite element (FE) analyses. Chinchalkar[18] also used the approximation of a spring with rotational stiffness within the conventional finite element method (FEM) to solve the free vibration problem for cracked wedges and two-segment beams. Lee[19] used the conventional FEM to analyse cantilever beams with two and three cracks. Therefore, it is necessary to reliably solve problems for the accurate frequencies and modes of non-uniform beams with multiple cracks. Furthermore, damage detection can be well developed based on these dynamic solutions. The above FE methods are generally not adaptivity-oriented and lack aspects required in an adaptive package.

To improve the validity and reliability of conventional FEM for solving free vibration problems, adaptive FEM has been proposed. Superconvergent patch recovery (SPR) and the corresponding adaptive technique originally proposed by Zienkiewicz and Zhu[20, 21] have been applied to static and dynamic problems to estimate spatial discretisation errors and to improve the solution of stresses. For problems on the free vibration of beams and structures without cracks, Wiberg et al. [22, 23] presented an application of local and global updating methods to improve the natural frequencies and modes predicted by the FE solutions in free vibration analysis, in which the local updating was based upon the superconvergent patch recovery displacements technique. The adaptive mesh refinement technique of FEM has been utilized to establish a three-dimensional model for rock stability analysis[24, 25]. Furthermore, adaptive FEM has been applied to successfully solve structure eigenvalue problems: e. g. buckling problems for non-uniform Euler-Bernoulli beam members[26] and free vibration problems for two-dimensional structures[27].

Considering the practical and theoretical importance of these problems, the crack detection problem has been extensively investigated as an inverse eigenproblem in structures, and many methods have been proposed to solve this problem, as have been comprehensively summarized in the summary of Dimarogonas[28]. Most fundamental studies concerning crack detection in a beam dealt with cases of single crack. The frequency contour plot method has been one of the most favoured tools to identify a single crack by using the lowest three natural frequencies obtained via a frequency measurement method[29, 30]. Owolabi et al. [31] proposed that the location and size of a crack could be identified by finding changes in frequencies and amplitudes of frequency response functions. A beam with multiple cracks was modelled as a massless rotational spring or other models based on the Euler-Bernoulli theory, and this scheme was subsequently adopted for crack detection in stepped beams[32-34]. In most studies, the crack was assumed to be open and normal to the beam surface. Morassi[35] studied crack detection problems involving an inclined-edge-type crack or a crack beneath the beam surface. Lele and Maiti[36] and Nikolakopoulos et al. [37] extended the frequency contour plot method to the crack detection in beams based on the Timoshenko beam theory and in-plane frame, respectively. In many cases, the three curves of the frequency contour plot unfortunately did not intersect because of the inaccuracy of modelling results as compared to measured results, and the zero-setting procedure was recommended for such cases[32-35]. Narkis[38] showed that, if a crack is very small, the only information required for crack detection is the variation of the first two natural frequencies due to a crack. Dado[39] presented a direct mathematical model to detect a crack in a beam, where the lowest two natural frequencies were required as input data. Hu and Liang[40] introduced a technique to detect multiple cracks. The continuum damage model was used to identify the discretizing elements of a structure that contained the cracks, and the spring damage model was used to quantify the location and size of the discrete crack in each damaged element. Patil and Maiti[41] presented a method that combined vibration modelling through the transfer matrix method and the approach proposed by Hu and Liang[40]. The detection of multiple cracks in beams was regarded as an optimization problem by Ruotolo and Surace[42], who selected the combination of fundamental functions as the objective function and utilized a solution procedure employing generic algorithms. Shifrin and Ruotolo[43] proposed that $n + 2$ equations are sufficient to form the system determinant for a beam with n cracks. Labib et al. used the Wittrick-Williams algorithm and dynamic stiffness method to analyse the free vibration and detect the cracks of cracked structures[10, 44], but their analytical method is only applicable to beams with uniform cross sections. Using the traditional FEM and cracks modelled as rotational springs, Chinchalkar[18] determined crack location in beams using natural frequencies for cracked wedge and two-segment beams, on the other hand, in the similar approach, Lee et al. [19] identified a cantilever beam with triple as well as double cracks. The above FE methods are generally not adaptivity oriented and lack the ingredients required in an adaptive package. Furthermore, the basic methodology for damage detection based on the dynamic solutions is well developed; on the other hand, it is necessary to solve the problem of acquiring reliable and accurate frequencies and modes for non-uniform beams with multiple cracks.

This chapter initially introduced an adaptive method based on the conventional FEM and the superconvergent patch recovery displacement technique for solving forward eigenproblems of beams with multiple cracks. The superconvergent computation technique is applied to calculate superconvergent solutions, which are henceforth referred to as superconvergent solutions, for eigenfunctions during the FE post-processing stage. These superconvergent solutions are then used as if they were exact solutions to estimate the errors in the FE solutions, which are used to guide mesh refinement. This yields a simple, efficient, reliable, and general adaptive FE procedure that can find sufficiently fine meshes to obtain FE solutions with the desired accuracy for the eigenvalues and eigenfunctions for beams with multiple cracks. Based on the solution dynamic solutions of forward eigenproblems, the inverse eigenproblems could be solved smoothly. The objective of the present study is to present an adaptive FE algorithm and procedure based on the adaptive FE solutions of frequencies via the superconvergent computation technique and on the Newton-Raphson iteration technique to identify multiple cracks in a beam, which requires $2n$ natural frequencies to detect n cracks in a beam. In this chapter, the presented procedure is applied to inverse eigenproblems of a beam with multiple cracks by utilizing the Newton-Raphson iteration technique to obtain damage information. Our simple, efficient, reliable, and generally adaptive FE procedure can find sufficiently fine meshes such that the obtained FE solutions satisfy the pre-specified error tolerance for both the locations and sizes of cracks in beams with multiple cracks.

5. 2   Adaptive approach for damage detection of cracked beams

5. 2. 1   Formulation and analogy of cracked beams

The goal of solving a regular fourth-order eigenproblem for a beam member based on Euler-Bernoulli beam theory is to find the frequencies ω and modes w of the fourth-order ordinary differential equation (ODE),

$$Lw \equiv (EI(x)w'')'' = \omega^2 m(x)w, \ 0 < x < l \tag{5.1}$$

subject to the default boundary conditions (BCs),

$$w(0)=0, \ w'(0)=0, \ w(l)=0, \ w'(l)=0 \tag{5.2}$$

where $EI(x)$ is the member's flexural rigidity, E is the elastic modulus, $m(x)$ is the linear density, and l is the length of the beam. The symbol L used in Equation (5.1) is the associated fourth-order self-adjoint operator. In the eigenproblem, the frequency and mode are the eigenvalue and eigenfunction, respectively, which together are known as an eigenpair.

Figure 5. 3 demonstrates a geometric model of a beam with cracks. Here, parameters $\alpha = a/h$ and $\beta = s/l$ denote the normalized crack depth and location, respectively, where a and s are the absolute crack depth and location, and h is the height of the beam.

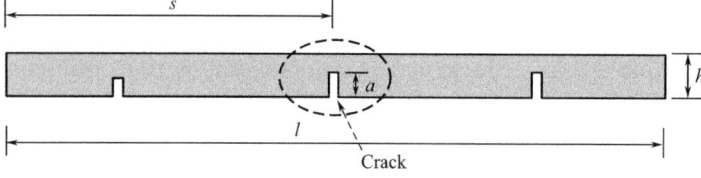

Figure 5. 3　Beam with multiple cracks

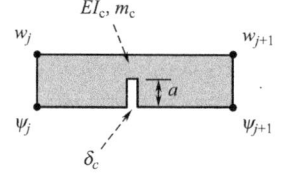

Figure 5. 4　FE model with crack

In the immediate region surrounding a single crack, the FE element containing the crack has two nodes with four degrees of bending and rotational freedoms (w_j, ψ_j) and (w_{j+1}, ψ_{j+1}), as shown in Figure 5. 4, where the narrow crack is described with a width δ_c set at $0.01 \times Tol$, where Tol is the pre-specified error tolerance for both frequencies and modes. Using the weakened properties analogy to reflect the presence of cracks, the flexural rigidity and density at the crack are reduced as the crack deepens:

$$EI_c = \frac{Ebh^3(1-\alpha)^3}{12} \tag{5.3a}$$

$$m_c = \overline{m}bh(1-\alpha) \tag{5.3b}$$

where EI_c and m_c are the flexural rigidity and linear density at crack c respectively; b is the width of the beam and $\overline{m}$ is the density.

5. 2. 2　Stop criterion

Suppose n cracks (α_i, β_i) $(i=1, 2, \cdots, n)$ are required and the pre-specified error tolerance for both the locations and sizes is Tol. The ultimate aim of the procedure presented here is to find FE solutions (α_i^h, β_i^h) $(i=1, 2, \cdots, n)$ on sufficiently fine meshes π such that

$$\alpha_i - \alpha_i^h < Tol \cdot (1+\alpha_i), \ i=1, 2, \cdots, n \tag{5.4a}$$

$$\beta_i - \beta_i^h < Tol \cdot (1+\beta_i), \ i=1, 2, \cdots, n \tag{5.4b}$$

Since the exact solutions (α_i, β_i) are not usually available, the proposed procedure uses the following stop criterion instead:

$$\omega_k^h - \omega_k < Tol \cdot (1+\omega_k), \ k=1, 2, \cdots, 2n \tag{5.5}$$

where ω_k and ω_k^h are the actual and computed frequencies of cracked beams, respectively. The above stop criteria in absolute error estimation for eigensolutions in adaptive analysis show satisfying effect[26, 27, 45].

In detail, as summarized in Table 5.1, the ultimate aim of damage detection is to obtain the exact solution of the problems. Unfortunately, the exact solution cannot be obtained for major problems; consequently, no solution can be used as the stop criterion. Therefore, the proposed procedure uses the new stop criterion introduced in Equation

(5.5), which has been shown to be effective through some numerical results involving the examples in Section 7. According to the uniqueness theory of the solution, the errors of each crack location and size compared to the exact solution are consistent with the errors of each frequency compared to the actual frequency for the structure with cracks; therefore, the latter is used as the stopping criterion in the proposed method.

Ultimate aim and stop criterion   **Table 5.1**

Problem type	Ultimate aim	Stop criterion
Damage detection problem (Inverse eigenproblem)	☒Errors of FE solutions of each crack location and size compared to the exact solution are less than *Tolerance*	☑Errors of FE solution of each frequency compared to the actual frequency for a beam with cracks are less than *Tolerance*

5.2.3　Analysis strategy

The adaptive FEM algorithm contains the free vibration analysis and damage detection of a beam with cracks, which are the forward eigenproblem and inverse eigenproblem, respectively, as shown in Figure 5.5. The proposed method intends to combine the free vibration analysis and the Newton-Raphson iteration technique to solve the inverse eigenproblem through the following three-step adaptive strategy:

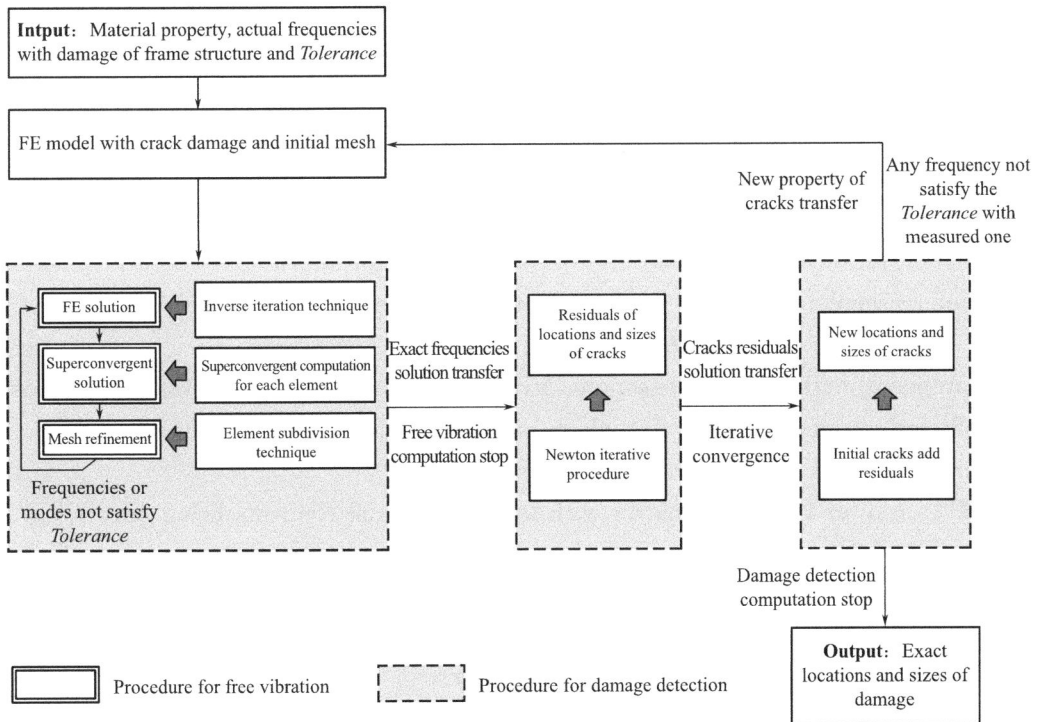

Figure 5.5   Adaptive FEM algorithm flowchart for free vibration problem (forward eigenproblem) and damage detection problem (inverse eigenproblem)

(1) Adaptive analysis. Under the current crack damage condition (the initial crack damage is provided by the user), the computed frequencies fully satisfy the pre-specified error tolerance by the adaptive FEM (forward eigenproblems), as described in Section 3.

(2) Newton-Raphson iteration. Utilizing the difference between computed frequencies and actual frequencies, the damage residuals of the computed and actual cracks are obtained by the Newton-Raphson iteration technique, as described in Section 4.

(3) Damage refinement. The crack locations and sizes are updated by the residuals of cracks damage to form the new crack damage condition. Then, the procedure returns to the first step (i. e. , adaptive analysis) until all frequency errors satisfy the pre-specified error tolerance, as described in Section 5.

5. 3   Adaptive analysis

5. 3. 1   Finite element solution

Utilizing the conventional FEM, the element stiffness matrix $\boldsymbol{K}^e$ and mass matrix $\boldsymbol{M}^e$ are computed and assembled to form the global stiffness matrix $\boldsymbol{K}$ and mass matrix $\boldsymbol{M}$. The FE equation of a beam member based on Euler-Bernoulli beam theory can be derived as an eigenvalue equation in the following matrix form:

$$\boldsymbol{KD} = \omega^2 \boldsymbol{MD} \tag{5.6}$$

where $\boldsymbol{D}$ is the mode vector, and the matrices $\boldsymbol{K}$ and $\boldsymbol{M}$ are both independent of ω. The element model adopted is the conventional polynomial element of degree $m > 3$, and $w^h \in C^1$ denotes the conventional FE solution on the given mesh π, in which C^1 is the space of functions that are continuous up to their first-order derivative. As in common practice, the shape functions for w^h are Hermite polynomials. Given an arbitrary trial value ω_a as the shift value, Equation (5. 6) can be equivalently written in the shifted form[46]

$$\boldsymbol{K}_a \boldsymbol{D} = \mu \boldsymbol{MD} \quad \text{with} \quad \boldsymbol{K}_a = \boldsymbol{K} - \omega_a^2 \boldsymbol{M}, \ \mu = \omega^2 - \omega_a^2 \tag{5.7}$$

In the proposed method, the convectional FE computation for eigenpair solutions is based on the Sturm sequence property[47], which can be expressed as

$$\boldsymbol{K} - \omega^2 \boldsymbol{M} = \boldsymbol{LD}(\omega) \boldsymbol{L}^{\mathrm{T}} \tag{5.8}$$

where $\boldsymbol{L}$ is a lower triangular matrix with leading diagonal elements being one, $\boldsymbol{L}^{\mathrm{T}}$ is its transpose, and $\boldsymbol{D}(\omega)$ is a diagonal matrix in which the number of eigenvalues less than the arbitrary trial value ω_a equals the number of negative leading diagonal elements in $\boldsymbol{D}(\omega_a)$. The Rayleigh quotient is used to accelerate the convergence on the eigenvalues:

$$\omega^2 = \frac{\boldsymbol{D}^{\mathrm{T}} \boldsymbol{KD}}{\boldsymbol{D}^{\mathrm{T}} \boldsymbol{MD}} \tag{5.9}$$

Utilizing the above Sturm sequence property and the convectionalbisection method[47], the intervals of each eigenvalue can be determined, and the inverse iteration technique is successfully introduced to compute the eigenpairs[25, 45]. Based on these considerations, the following inverse iteration procedure is adopted:

$$\begin{cases} \overline{\boldsymbol{D}}_{i+1} = \boldsymbol{K}_a^{-1} \boldsymbol{M} \boldsymbol{D}_i \\ \mu_{i+1} = \dfrac{\overline{\boldsymbol{D}}_{i+1}^{\mathrm{T}} \boldsymbol{M} \boldsymbol{D}_i}{\overline{\boldsymbol{D}}_{i+1}^{\mathrm{T}} \boldsymbol{M} \overline{\boldsymbol{D}}_{i+1}} \\ \boldsymbol{D}_{i+1} = \mathrm{sgn}(\mu_{i+1}) \dfrac{\overline{\boldsymbol{D}}_{i+1}}{\max(\overline{\boldsymbol{D}}_{i+1})} \end{cases} \tag{5.10}$$

where i is the loop index. The above inverse iteration procedure is terminated when the following conditions are met

$$|\mu_{i+1} - \mu_i| < Tol \text{ and } \max|\boldsymbol{D}_{i+1}| < Tol \tag{5.11}$$

After the above inverse iteration converges, an FE solution $(\mu^h, \boldsymbol{D}^h)$ (i. e. $(\omega^h, \boldsymbol{D}^h)$ where $(\omega^h)^2 = \omega_a^2 + \mu^h$) is obtained. However, the current mesh may not be sufficiently fine, in which case the accuracy of this FE solution must be estimated by a more accurate solution, namely, the superconvergent solution, which is discussed in the following section.

5.3.2   Error estimation and mesh refinement

The superconvergent patch recovery displacement technique was developed for computation of superconvergent displacements for FE solutions of static and dynamic problems[22, 23]. The displacements provided by the superconvergent computation technique can be applied to eigenfunctions. For example, as shown in Figure 5.6, element e is the superconvergent computation element, and elements e-1 and $e+1$ are neighbouring elements, in which FE nodes j-1, j, $j+1$, and $j+2$ are selected for computation.

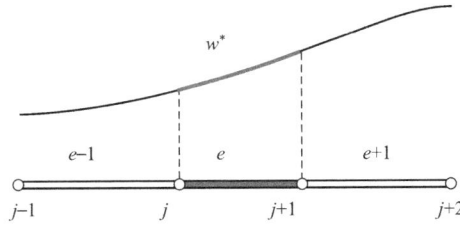

Figure 5.6   Computation of superconvergent displacements for element e

The superconvergent displacements for element e can be computed as

$$w^*(x) = \sum_{i=1}^r N_i(x) w_i^h + \sum_{i=1}^s N_i(x) \overline{w}_i^* \tag{5.12}$$

where r ($=2$) is the number of end nodes, s is the number of internal nodes, and $N_i(x)$ is the shape function. The degree of the shape function is improved by one order as $r+s = m+1$. To make the best use of the superconvergent order $O(h^{2m})$ for displacements at end nodes, the displacement recovery field can be expressed by FE nodes as

$$\overline{w}^*(x) = \boldsymbol{Pa} \tag{5.13}$$

where $\boldsymbol{P}$ is the given function vector, and $\boldsymbol{a}$ can be obtained by the least squares fitting technique for the coincidence of displacements at the end nodes in the recovery field and the conventional FE field. The superconvergent displacements at the end nodes in recovery field $\overline{w}^*(x)$ are used in Equation (5.12) to obtain the superconvergent solutions on element e. Because the accuracy of the superconvergent solution w^* is at least one order higher than that of w^h, for elements of degree $m > 3$, a very simple strategy for error estimation is to use w^* instead of the exact solution w to estimate the errors in w^h. This error estimation method has shown good reliability and effectiveness[22, 23].

The superconvergent solutions of Equation (5.13) can be used in the Rayleigh quotient[47] to obtain estimates of the eigenvalue:

$$\omega_k^* = \sqrt{\frac{a(w_k^*, w_k^*)}{b(w_k^*, w_k^*)}}, \ k = 1, 2, \cdots, 2n \tag{5.14}$$

where $a(w, v) = \int_0^l EI(x) w'' v'' \mathrm{d}x$ and $b(w, v) = \int_0^l m(x) wv \mathrm{d}x$ are the strain energy inner product and the kinematic energy inner product, respectively. The estimated eigenvalue is a stationary value when taken over all possible functions that satisfy the essential BCs. The stationary values computed by Equation (5.14) are superconvergent eigenvalues and the corresponding functions ω_k^* are the superconvergent solutions. The Rayleigh quotient, Equation (5.14), can be expressed based on elements as

$$\omega_k^* = \sqrt{\frac{\sum_e \int_{a_e}^{b_e} EI(x) w_k^{*''} w_k^{*''} \mathrm{d}x}{\sum_e \int_{a_e}^{b_e} m(x) w_k w_k \mathrm{d}x}}, \ k = 1, 2, \cdots, 2n \tag{5.15}$$

where a_e and b_e are the end nodes of the boundary for element e.

Here, each element on the current mesh is divided into a grid of M equal subintervals. For the $M-1$ interior grid points on a typical element e, the conventional FE solutions w_g^h and the superconvergent solutions w_g^* at the g-th interior point ($g = 1, 2, \cdots, M-1$) are calculated. Then, the errors at the $M-1$ interior points are calculated and estimated to determine if all of them satisfy the given tolerance:

$$\| e_{w,k}^* \|_e \leqslant Tol \cdot [(\| w_k^h \|^2 + \| e_{w,k}^* \|^2)/n]^{1/2}, \ k = 1, 2, \cdots, 2n \tag{5.16}$$

where $e_{w,k}^*$ is the error of the superconvergent displacements w_k^* and the computed displacements w_k^h, n_e is the number of elements, $\| w \| = [a(w, w)]^{1/2}$. Equation (5.16) can be equivalently written in the following form

$$\xi_k = \frac{\| e_{w,k}^* \|_e}{\overline{e}_{w,k}}$$

with $\overline{e}_{w,k} = Tol \cdot [(\| w_k^h \|^2 + \| e_{w,k}^* \|^2)/n_e]^{1/2}$ \qquad $k = 1, 2, \cdots, 2n$ \tag{5.17}

where ξ_k should satisfy

$$\xi_k \leqslant 1, \ k = 1, 2, \cdots, 2n \tag{5.18}$$

Usually it is more than sufficient to set M in the range of $4 \leqslant M \leqslant 8$. Therefore, without loss of generality, M is set to 6 for the remainder of this chapter.

If Equation (5.18) is not satisfied for any interior point, the corresponding element needs to be subdivided into uniform sub-elements by inserting some interior nodes through h-refinement[20, 21, 46], which are calculated by

$$h_{k,\text{ new}} = \xi_k^{-1/m} h_{k,\text{ old}}, \quad k=1, 2, \cdots, 2n \tag{5.19}$$

where $h_{k,\text{ new}}$ is the length of the sub-element, $h_{k,\text{ old}}$ is the original length of element e, and $\lfloor \cdot \rfloor$ represents the "floor" operator, i. e. rounding down to the nearest integer. The above element subdivision approach is implemented as

$$n_{k,\text{ new}} = \min(\lfloor \xi_k^{-1/m} \rfloor, d), \quad k=1, 2, \cdots, 2n \tag{5.20}$$

where $n_{k,\text{ new}}$ is the number of subelements after element subdivision, and d is the limit number for avoiding too many redundant elements. Each element e that does not satisfy the pre-specified error tolerance is uniformly subdivided, e. g. $h_{k,\text{ new}} = h_{k,\text{ old}}/6$ as shown in Figure 5.7.

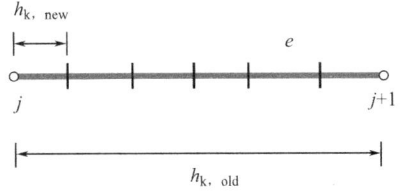

Figure 5.7　Uniform subdivision of element e (e. g. $h_{k,\text{ new}} = h_{k,\text{ old}}/6$)

5.4　Newton-Raphson iteration

Based on the frequency measurement method, the residuals of the frequencies and cracks are consistent with each other; therefore, the Newton-Raphson iteration technique can be introduced[47]. For the identification of n cracks in a beam, there should be $2n$ unknown crack parameters: α_1, β_1, α_2, β_2, $\cdots$, α_n, and β_n. To match the number of equations and the number of unknown parameters, it is assumed that $2n$ natural frequency measurements ω_1^0, ω_2^0, $\cdots$, ω_{2n-1}^0 and ω_{2n}^0 are available in advance. The Newton-Raphson iteration procedure is applied in this chapter as follows:

(a) The user assumes initial values of α_1, β_1, α_2, β_2, $\cdots$, α_n, and β_n and the FE mesh of the beam.

(b) Locate the positions that represent the cracks according to the new crack locations parameters β_1, β_2, $\cdots$, β_n.

(c) Solve the forward eigenproblem to obtain FE solutions for ω_1^h, ω_2^h, $\cdots$, ω_{2n}^h with the crack parameters α_1, β_1, α_2, β_2, $\cdots$, α_n, and β_n, and evaluate the Jacobian matrix

$$\boldsymbol{J} = \begin{bmatrix} \dfrac{\partial \omega_1^h}{\partial \alpha_1} & \dfrac{\partial \omega_1^h}{\partial \beta_1} & \dfrac{\partial \omega_1^h}{\partial \alpha_2} & \dfrac{\partial \omega_1^h}{\partial \beta_2} & \cdots & \dfrac{\partial \omega_1^h}{\partial \alpha_n} & \dfrac{\partial \omega_1^h}{\partial \beta_n} \\[2ex] \dfrac{\partial \omega_2^h}{\partial \alpha_1} & \dfrac{\partial \omega_2^h}{\partial \beta_1} & \dfrac{\partial \omega_2^h}{\partial \alpha_2} & \dfrac{\partial \omega_2^h}{\partial \beta_2} & \cdots & \dfrac{\partial \omega_2^h}{\partial \alpha_n} & \dfrac{\partial \omega_2^h}{\partial \beta_n} \\[2ex] \vdots & \vdots & \vdots & \vdots & & \vdots & \vdots \\[2ex] \dfrac{\partial \omega_{2n}^h}{\partial \alpha_1} & \dfrac{\partial \omega_{2n}^h}{\partial \beta_1} & \dfrac{\partial \omega_{2n}^h}{\partial \alpha_2} & \dfrac{\partial \omega_{2n}^h}{\partial \beta_2} & \cdots & \dfrac{\partial \omega_{2n}^h}{\partial \alpha_n} & \dfrac{\partial \omega_{2n}^h}{\partial \beta_n} \end{bmatrix} \tag{5.21}$$

and compute the residuals of frequencies

$$R_k = \omega_k^h - \omega_k^0, \quad k = 1, 2, \cdots, 2n \tag{5.22}$$

(d) Solve the following equation by Newton-Raphson iteration:

$$\boldsymbol{J}\, \mathrm{d}\boldsymbol{C}^{\mathrm{T}} = -\boldsymbol{R}^{\mathrm{T}} \tag{5.23}$$

where $\boldsymbol{C}^{\mathrm{T}} = (\alpha_1, \beta_1, \alpha_2, \beta_2, \cdots, \alpha_n, \beta_n)^{\mathrm{T}}$ and $\boldsymbol{R}^{\mathrm{T}} = (R_1, R_2, \cdots, R_n)^{\mathrm{T}}$, through which the residuals of n cracks $(\mathrm{d}\alpha_i, \mathrm{d}\beta_i)(i = 1, 2, \cdots, n)$ will be obtained.

(e) Update the crack parameters by utilizing the residuals of cracks:

$$(\alpha_i)_{\mathrm{new}} = (\alpha_i)_{\mathrm{old}} + \mathrm{d}\alpha_i, \quad (\beta_i)_{\mathrm{new}} = (\beta_i)_{\mathrm{old}} + \mathrm{d}\beta_i, \quad i = 1, 2, \cdots, n \tag{5.24}$$

where $()_{\mathrm{new}}$ and $()_{\mathrm{old}}$ represent the new and old cracks in the last step, respectively. Update each old crack with the new one, as shown in Figure 5.8.

(f) In the new crack condition, return to step (a) and repeat the loop until the residuals of frequencies become sufficiently small.

Note that the FE mesh of conventional FEM for the Newton-Raphson iteration procedure is determinate without mesh refinement. However, the adaptive FE analysis proposed in this chapter will have more accurate results and a better convergence rate compared to the conventional FE analysis, which is shown in the numerical examples in Section 7.

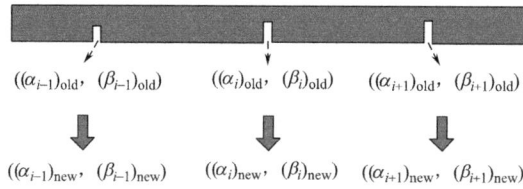

Figure 5.8 Update of size and location of cracks in one iteration step

5.5 Damage refinement

The matrix elements in the Jacobian matrix $\boldsymbol{J}$ are related to the frequencies and crack parameters because the most widely used method, damage detection based on the optimization theory, is reduced to a linearized system of equations. The matrix elements of the Jacobian matrix $\boldsymbol{J}$ are the sensitivities of the natural frequencies with respect to the crack parameters. Morassi[35] developed an explicit expression of the frequency sensitivity to damage, assuming that the sizes of the cracks were sufficiently small. In this study, however, the cracks are not assumed to be small, and the elements of the Jacobian matrix $\boldsymbol{J}$ are computed numerically introducing the method by Lee[19]. For example, $\partial\omega_1/\partial\alpha_1$ and $\partial\omega_1/\partial\beta_1$ are computed, respectively, as follows:

$$\frac{\partial\omega_1}{\partial\alpha_1} = \frac{\omega_1(\alpha_1 + \delta, \beta_1, \alpha_2, \cdots, \beta_n) - \omega_1(\alpha_1, \beta_1, \alpha_2, \cdots, \beta_n)}{\delta}, \quad |\delta| \ll 1 \tag{5.25a}$$

$$\frac{\partial\omega_1}{\partial\beta_1} = \frac{\omega_1(\alpha_1, \beta_1 + \delta, \alpha_2, \cdots, \beta_n) - \omega_1(\alpha_1, \beta_1, \alpha_2, \cdots, \beta_n)}{\delta}, \quad |\delta| \ll 1 \tag{5.25b}$$

where δ is a value far less than one. Because the crack condition is inaccurate in the initial stage of Newton-Raphson iteration, the residuals of cracks can be adjusted to accelerate the procedure (Lee, 2009). The forward eigenproblem is solved $2n+1$ times per iteration to build the Jacobian matrix $\boldsymbol{J}$ and the residuals. To suppress overshoots in the early stage, relaxation is performed during the beginning iterations steps as follows:

$$(\alpha_i)_{new} = (\alpha_i)_{old} + 0.25d\alpha_i, \quad i = 1, 2, \cdots, n \tag{5.26a}$$

$$(\beta_i)_{new} = (\beta_i)_{old} + 0.25d\beta_i, \quad i = 1, 2, \cdots, n \tag{5.26b}$$

5.6 Algorithms

The basic algorithm of the proposed adaptive FEM for free vibration problems (forward eigenproblems) is given as follows:

(1) For the kth order eigenpair (frequency and mode), the initial FE mesh is imported from the final mesh for the previous eigenpair solution, $\pi_{k-1}(k=1, 2, \cdots, 2n)$ (the initial mesh π_0 for the first-order eigenpair and the pre-specified error tolerance Tol are given by user).

(2) For the current order and adaptive step, the FE mesh π_k^t is obtained.

(3) The material properties (i.e. flexural rigidity andlinear density) are reduced at the cracks using Equation (5.3) to form the whole beam model.

(4) The conventional FE solutions for the eigenpair $(\omega_k^h, w_k^h)(k=1, 2, \cdots, 2n)$ are computed on the current mesh π_k^t utilizing the inverse iteration procedure of Equation (5.10).

(5) Superconvergent FE solutions for eigenpair $(\omega_k^*, w_k^*)(k=1, 2, \cdots, 2n)$ are computed using the superconvergent patch recovery displacement methodology and Rayleigh quotient of Equations (5.12) and (5.14), respectively.

(6) The errors of the FE solutions are estimated utilising thesuperconvergent solutions, in the implementation of the procedure, these errors are estimated at interior points using Equation (5.18); if the errors are not satisfied, element subdivision Equation (5.20) is used to form the new FE mesh π_k^{t+1}.

(7) The mesh index is updated as $t=t+1$ and the algorithm returns to step (2) unless the errors are satisfied.

(8) The kth order mesh is finalized as $\pi_k = \pi_k^t$, the eigenpair index is updated $k=k+1$, and the algorithm returns to step (1) unless the final order $(k=2n)$ has finished.

Based on the computation of free vibration problems (forwardeigenproblems), the global algorithm of the proposed adaptive FEM for damage detection problems (inverse eigenproblems) is proposed as follows:

(1) The actual frequencies of the beam with cracks and initial predicted cracks c_0 are provided, noting that the detection for n cracks needs $2n$ actual frequencies. The initial FE mesh π_0 and the pre-specified error tolerance Tol are given by the user.

(2) Introducing the above basic algorithm for free vibration problems, theconventional FE solutions for eigenpair $(\omega_k^h, w_k^h)(k = 1, 2, \cdots, 2n)$ and superconvergent FE solutions for eigenpair $(\omega_k^*, w_k^*)(k = 1, 2, \cdots, 2n)$ on the current mesh are computed. Then the errors of the FE solutions are estimated and the element subdivisions are evolved until the errors are satisfied based on the mesh refinement. Finally, accurate and reliable frequencies of the current crack condition are obtained.

(3) The actual and computed frequencies are compared to analyse their residuals using Equation (5. 22).

(4) With the Newton-Raphson iteration technique and frequency residuals, the crack residuals are computed using Equation (5. 23).

(5) With the current crack condition and crack residuals, the new crack condition is computed using Equation (5. 24).

(6) In the new crack condition, return to step (1) and repeat the loop until the stop criterion in Equation (5. 5) is satisfied.

5. 7 Numerical examples

The proposed adaptive strategy has been coded into a Fortran 90 program; in this section, it was verified by solving four representative numerical examples for free vibration problems, on the other hand, which examples are also selected as damage detection problems to show that the method is correct and competitive. Also, for comparison, whenever needed, both the free vibration and damage detection problems are dealt with together by setting one-to-one corresponding examples respectively, e. g. Example 1 and Example 5 cases. All of these examples were run utilizing the Intel (R) Visual Fortran Compiler on a DELL Optiplex 380 desktop computer with an Intel (R) Core (TM) 2. 93 GHz CPU, with double-precision floating-point numbers (approximately 14 decimal digits). For all the examples, the tolerance Tol is set to 10^{-3} and the fifth-order ($m = 5$) polynomials are used for each element.

For free vibration problems, the error of the computed frequency ω^h is

$$\varepsilon_\omega = \frac{|\omega - \omega^h|}{1 + |\omega|} \tag{5. 27}$$

where ω is the exact frequency or a reliable result obtained through other methods. The first example is a double-clamped non-uniform uncracked beam, whose exact solution can be computed using SLUTH[13] with a strict tolerance setting of 10^{-9}. Therefore, the results obtained from the present method are compared with these exact solutions. For the other three examples, only the calculated results from other studies are available because exact solutions are not available for cracked beams. Every eigenfunction solution shown below is normalized to its biggest value.

For damage detection, the errors of the computed crack depth α^h and location β^h are

$$\varepsilon_\alpha = \frac{|\alpha - \alpha^h|}{1 + |\alpha|}, \quad \varepsilon_\beta = \frac{|\beta - \beta^h|}{1 + |\beta|} \tag{5.28}$$

where α and β are the exact crack depth and location or reliable results obtained through other methods. For all the examples, it was found that the present procedure produced satisfactory results, with both locations and sizes of cracks fully satisfying the pre-specified error tolerance.

5.7.1　Example 1: Double-clamped uncracked beam with a sinusoidal cross section

Figure 5.9 (a) shows the double-clamped uncracked beam, and the material data are

$$h(x) = h_0\left(1 + 0.5\sin\left(10\frac{x}{l}\right)\right), \quad m(x) = \rho b h(x), \quad EI(x) = \frac{Ebh^3(x)}{12} \tag{5.29}$$

This example was selected to check the reliability of the proposed method for non-uniformuncracked beams. The first ten frequencies and the final adaptive mesh computed by the proposed method are shown in Table 5.2. This Table also displays the solutions computed using SLEUTH with a strict tolerance setting of 10^{-9}, which serves as the exact solution because this problem does not have an analytic solution. It is evident that the pre-specified error tolerance is well satisfied for frequencies. The first three computed modes are shown in Figure 5.9 (b); clearly, the vibration becomes more complicated as the order increases, which means that more elements are necessary to effectively analyse the free vibration problem, as shown in Table 5.2.

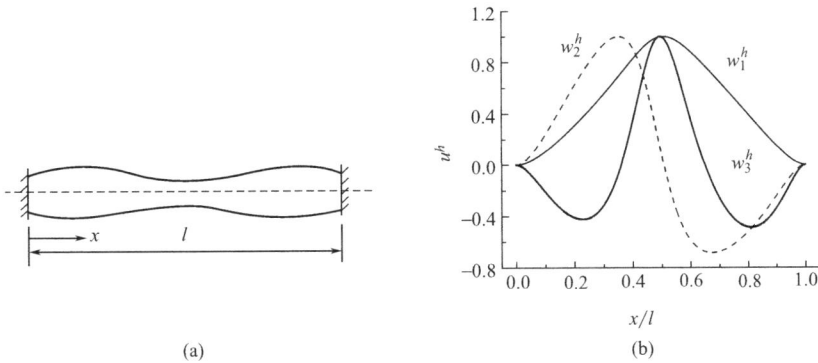

Figure 5.9　Model and modes of Example 1. Notes: (a) Double-clamped uncracked beam with a sinusoidal cross section. (b) Computed results for first three modes

Computed results for frequencies of Example 1			Table 5.2
k	Present method			SLEUTH solutions
	ω_k^h	ε_ω	Elements	ω_k
1	21.446122	7.97E−04	7	21.428254
2	56.805177	2.84E−05	9	56.803537
3	124.219103	6.28E−05	9	124.211235

continued

k	Present method			SLEUTH solutions
	ω_k^h	ε_ω	Elements	ω_k
4	196. 270778	5. 59E−04	9	196. 160646
5	293. 647288	2. 49E−05	12	293. 639963
6	411. 989517	1. 30E−05	12	411. 984134
7	550. 268718	2. 21E−05	12	550. 256549
8	708. 331094	9. 42E−06	16	708. 324415
9	886. 135673	5. 20E−06	16	886. 131057
10	1083. 668240	4. 66E−06	20	1083. 663183

5. 7. 2　Example 2: Stepped cantilever beam with a single crack

Figure 5. 10 (a) shows the stepped cantilever beam with a single crack with lengths l_1 and l_2, heights h_1 and h_2, and the following material data

$$l_1 = l_2 = 0. 25 \text{ m}, \quad E = 2. 1 \times 10^{11} \text{ Pa}, \quad \rho = 7800 \text{ kg/m}^3, \quad \nu = 0. 3 \quad (5. 30)$$

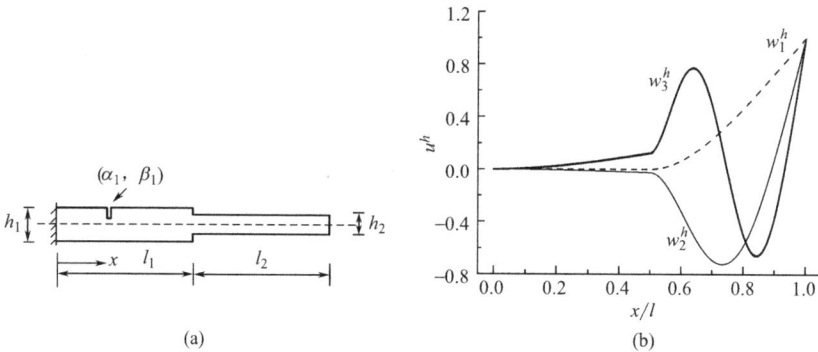

Figure 5. 10　Model and modes of Example 2. Notes: (a) Stepped cantilever beam with single crack.
(b) Computed results for first three modes of Case (c)

This model was previously analysed using a semi-analytical method[11]. Three cases are considered, as shown in Table 5. 3, for different sizes of cracks and different beam heights. The first four computed frequencies are shown in Table 5. 3. Therein, the results of the present method are compared with the solutions with the semi-analytical method for $h_1 = 0. 02$ m and $h_2 = 0. 16$ m as Cases (a) and (b). Furthermore, the frequencies were computed for a large step ratio $h_1 = 0. 02$ m and $h_2 = 0. 16$ m in Case (c) to check the stability of the proposed method; the exact solution for Case (c) was calculated using the proposed method with a strict tolerance setting of 10^{-9}. The computed solutions are consistent with the semi-analytical solutions for Cases (a) and (b); furthermore, the computed solutions for Case (c) using the pre-specified error tolerance $Tol = 10^{-3}$ agree with the solutions using the strict tolerance of 10^{-9}. The first three computed modes of

Case (c) are shown in Figure 5. 10 (b). This Figure 5. 11. shows that the modes become more complicated in the domain of relatively small stiffness. For further consideration, the 10th order modes for Cases (a) and (c) are shown in Figures. 5. 11 (a) and (b), respectively. In these figures, the final adaptive mesh is shown as tick marks on the horizontal axis. Because the stiffness is fairly constant in Case (a), the modes are smooth and the mesh is fairly uniform. In contrast, because the left half of the beam in Case (c) is much stiffer, the mode varies significantly throughout the beam, and a finer mesh is required near the tip of the beam.

Computed results for frequencies of Example 2　　　　　　**Table 5. 3**

Cracks		k	Present method		Results[1] (Cases (a) & (b))
			ω_k^h	ε_ω	ω_k
Case (a)	$\alpha_1=0.100$	1	457.308321	9.49E−03	453.0
	$\beta_1=0.200$	2	2376.679751	1.32E−02	2345.7
	$h_1=0.02\text{m}$	3	6649.213179	2.32E−02	6498.4
	$h_2=0.016\text{m}$	4	12790.812926	—	—
Case (b)	$\alpha_1=0.200$	1	457.308321	4.24E−02	477.6
	$\beta_1=0.200$	2	2376.679751	1.37E−02	2344.6
	$h_1=0.02\text{m}$	3	6649.213179	2.60E−02	6480.9
	$h_2=0.016\text{m}$	4	12790.812926	—	—
Case (c)	$\alpha_1=0.100$	1	1344.066976	4.97E−04	1344.736216
	$\beta_1=0.200$	2	8300.314483	1.44E−04	8301.508858
	$h_1=0.02\text{m}$	3	16333.267628	6.07E−04	16323.357462
	$h_2=0.016\text{m}$	4	24216.911236	1.31E−04	24220.074496

Source: ① Results from paper [11].

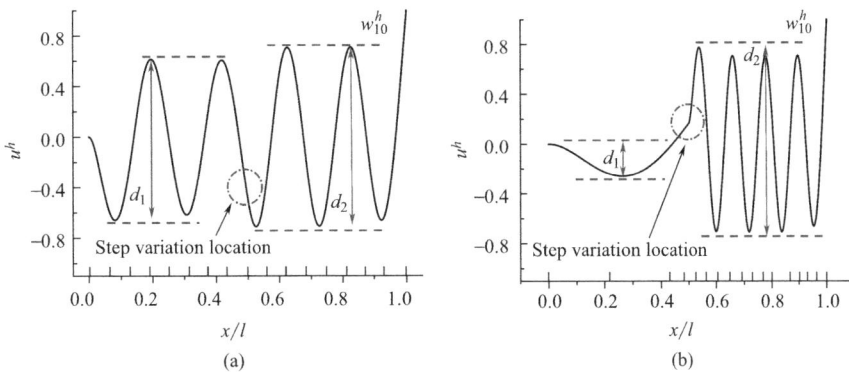

Figure 5. 11　Computed 10th order modes for different stepped height ratios and final meshes in Example 2. Notes: (a) Case (a). (b) Case (c)

5.7.3　Example 3: Cantilever beam with double cracks

Figure 5.12 (a) shows a cantilever beam with two cracks, and the material data are

$$l = 0.5 \mathrm{m}, \quad h = 0.02 \mathrm{m}, \quad E = 2.1 \times 10^{11} \mathrm{Pa}, \quad \rho = 7860 \mathrm{kg/m^3}, \quad \nu = 0.3 \quad (5.31)$$

This example was selected to check the reliability of the proposed method for computing the frequencies and corresponding modes of beams with multiple cracks. This model was previously analysed using the torsional spring method for simulating the cracks[19]. Three cases were considered, as shown in Table 5.4 for different locations and sizes of cracks. The first four computed frequencies are shown in Table 5.4. These solutions are compared with the solutions from the torsional spring method. The computed solutions are consistent with the torsional spring method for the three cases. The first three computed modes for Case (c) are shown in Figure 5.12 (b).

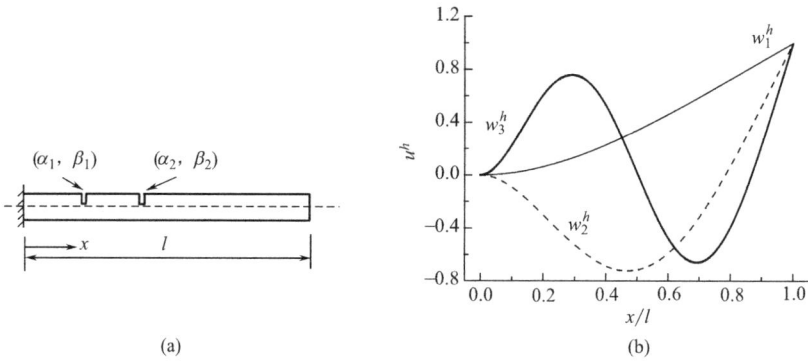

Figure 5.12　Model and modes of Example 3. Notes: (a) Cantilever beam with two cracks.
(b) Computed results for the first three modes of Case (c)

Computed results for frequencies of Example 3　　　　　　　　Table 5.4

Cracks		k	Present method		Results[①]
			ω_k^h	ε_ω	ω_k
Case (a)	$\alpha_1 = 0.1$	1	419.709187	6.29E−03	417.0794
	$\beta_1 = 0.2$	2	2630.272577	3.01E−03	2622.389
	$\alpha_2 = 0.1$	3	7364.839379	3.20E−03	7341.322
	$\beta_2 = 0.4$	4	14432.145781	4.45E−03	14368.22
Case (b)	$\alpha_1 = 0.1$	1	419.709187	3.56E−03	418.2175
	$\beta_1 = 0.4$	2	2630.272577	1.82E−02	2583.284
	$\alpha_2 = 0.2$	3	7364.839379	1.09E−02	7285.600
	$\beta_2 = 0.6$	4	14432.145781	4.02E−03	14374.36

continued

Cracks		k	Present method		Results[①]
			ω_k^h	ε_ω	ω_k
Case (c)	$\alpha_1=0.1$	1	419.709187	5.86E−04	419.4628
	$\beta_1=0.6$	2	2630.272577	6.99E−03	2612.009
	$\alpha_2=0.2$	3	7364.839379	1.43E−02	7260.865
	$\beta_2=0.8$	4	14432.145781	1.94E−02	14158.15

Source: ① Results from paper [19]

5.7.4　Example 4: Cantilever beam with triple cracks

Figure 5.13 (a) shows a cantilever beam with three cracks; the material data are the same as Example 3. Three cases were considered as shown in Table 5.5 for different locations and sizes of cracks. The first six computed frequencies are shown in Table 5.5. These solutions are compared with the solutions from the torsional spring method for Cases (a) and (b). Similar to Example 2, exact solutions were obtained for Case (c) using the proposed method with a strict tolerance of 10^{-9}. In Case (c), the frequencies were computed for deep cracks to check the stability of the proposed method. The computed solutions are consistent with the torsional spring method for Cases (a) and (b), and the solutions computed with the pre-specified error tolerance $Tol = 10^{-3}$ match the solutions with the tolerance of 10^{-9}. The first three computed modes of the deep cracks for Case (c) are shown in Figure 5.13 (b). This Figure 5. shows that the modes become more complicated as the cracks deepen. For further consideration, uncracked and deeply cracked beams are compared in Figures 5.14 (a) and (b), respectively. In these figures, the final adaptive mesh is shown as tick marks on the horizontal axis. Because there are no cracks in the case shown in Figure 5.14 (a), the modes are smooth and the mesh is fairly uniform. In contrast, the cracks in the case shown in Figure 5.14 (b) cause the mode to grow in magnitude as it passes through the cracks, and a finer mesh is necessary in the vicinity of the cracks.

Computed results for frequencies of Example 4　　　　　　**Table 5.5**

Cracks		k	Present method		Results[a]
			ω_k^h	ε_ω	ω_k
Case (a)	$\alpha_1=0.1$	1	419.709187	6.74E−03	416.8933
	$\beta_1=0.2$	2	2630.272577	6.97E−03	2612.065
	$\alpha_2=0.1$	3	7364.839379	5.59E−03	7323.879
	$\beta_2=0.4$	4	14423.380407	4.65E−03	14356.68
	$\alpha_3=0.1$	5	23823.591113	9.91E−03	23589.91
	$\beta_3=0.6$	6	35634.541339	8.59E−04	35603.94

continued

Cracks		k	Present method		Results[a]
			ω_k^h	ε_ω	ω_k
Case (b)	$\alpha_1 = 0.1$	1	419.709187	6.32E−03	417.0652
	$\beta_1 = 0.2$	2	2630.272577	3.78E−03	2620.375
	$\alpha_2 = 0.1$	3	7364.839379	6.34E−03	7318.436
	$\beta_2 = 0.4$	4	14415.115577	8.05E−03	14299.97
	$\alpha_3 = 0.1$	5	23824.135951	9.48E−03	23600.29
	$\beta_3 = 0.8$	6	35630.346532	1.59E−03	35573.62
Case (c)	$\alpha_1 = 0.3$	1	419.809187	2.38E−04	417.332753
	$\beta_1 = 0.2$	2	2630.309132	1.39E−05	2616.921931
	$\alpha_2 = 0.3$	3	7364.539463	4.07E−05	7335.876113
	$\beta_2 = 0.4$	4	14432.168001	1.54E−06	14377.867894
	$\alpha_3 = 0.3$	5	23657.374210	4.45E−05	23658.501681
	$\beta_3 = 0.6$	6	35638.841304	7.15E−04	35613.380555

Source: ① Results from paper Lee [19].

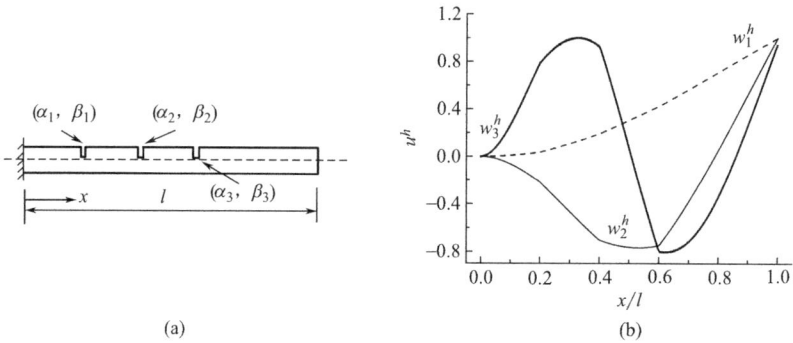

(a) (b)

Figure 5.13   Model and modes of Example 4. Notes: (a) Cantilever beam with three cracks. (b) Computed results for first three modes of Case (c)

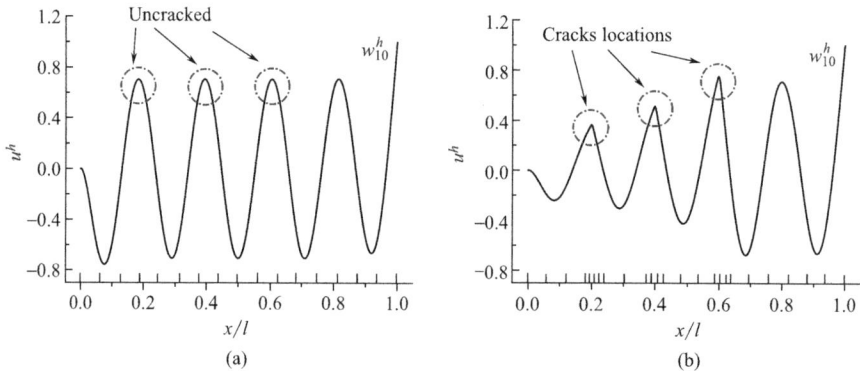

(a) (b)

Figure 5.14   Tenth-order computed modes and final meshes for different crack sizes in Example 4. Notes: (a) Uncracked beam: $\alpha_i = 0.0 (i = 1, 2, 3)$. (b) Deeply cracked beam: $\alpha_i = 0.3$ $(i = 1, 2, 3)$

5. 7. 5 Example 5: Double-clamped uncracked beam with a sinusoidal cross section

Consider the clamped-clampeduncracked beam in Figure 5. 15 (a) and the material data are the same as Example 1 as shown in Equation (5. 29). This example is selected for checking the reliability of the proposed method for non-uniform uncracked beams. Because exact frequencies do not exist in this case, the first four frequencies of forward eigenproblems computed by the proposed method are used as actual frequencies, as listed in Table 5. 6. Assuming the existence of two cracks, the first four frequencies and crack properties computed by the proposed method are listed in Table 5. 6. It can be seen that the two detected cracks are located at $\beta_1 = 0. 15224$ and $\beta_2 = 0. 65163$, and the differences between the computed and actual sizes (0. 000) of the cracks satisfy the pre-specified error tolerance $Tol = 10^{-3}$, revealing that there are no cracks in this non-uniform beam. In Table 5. 6, the differences between the computed and actual frequencies also satisfy the pre-specified error tolerance. The Newton-Raphson iteration results for cracks are shown in Figure 5. 15 (b), where the Newton-Raphson iteration converges after only 6 iteration steps.

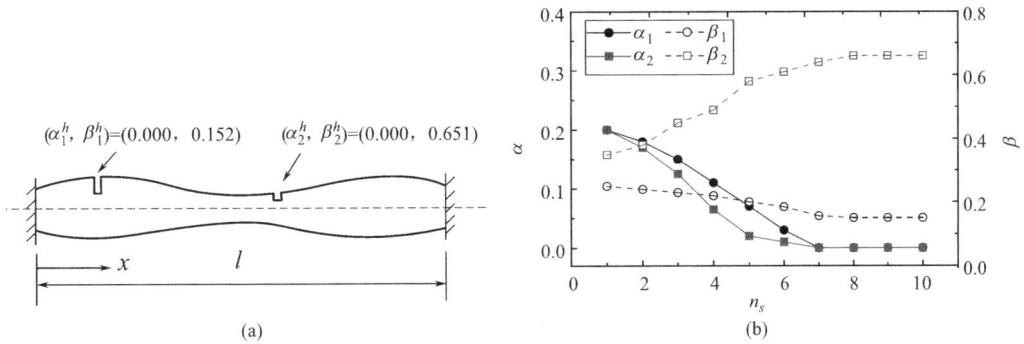

Figure 5. 15 Model and damage detection results of Example 5. Notes: (a) Clamped-clamped beam with sinusoidal cross section. (b) Newton-Raphson iteration results for cracks

5. 7. 6 Example 6: Stepped cantilever beam with a single crack

Consider the stepped cantilever beam witha single crack in Figure 5. 16 (a) with lengths l_1 and l_2, heights h_1 and h_2, and the material data are the same as Example 2 as shown in Equation (5. 30). This model had been analysed by a semi-analytical method[11]. Considering three cases with different locations and sizes of cracks as $\alpha_1 = 0. 100$, $\beta_1 = 0. 200$ and $\alpha_1 = 0. 200$, $\beta_1 = 0. 200$, the first four computed frequencies are listed in Table 5. 7 and compared with the solutions obtained using the semi-analytical method for $h_1 = 0. 02\text{m}$ and $h_2 = 0. 16\text{m}$ as case (a) and case (b). On the other hand, the frequencies were computed for a large step ratio with $h_1 = 0. 20\text{m}$ and $h_2 = 0. 16\text{m}$ as case (c). The first four frequencies of forward eigenproblems computed by the proposed method, considered as the actual frequencies, are listed in Table 5. 7. Assuming the existence of two cracks, the first four frequencies and crack properties computed by the proposed method are also listed in Table 5. 7. It can be seen that the two detected cracks are located at $\alpha_1 = 0.100$, $\beta_1 = 0.200$ and $\alpha_1 = 0.200$, $\beta_1 = 0.200$, and the differences between the computed and actual sizes of the cracks satisfy the pre-specified error tolerance, revealing that there is only one crack in this non-uniform beam. In Table 5. 7, the differences between computed and actual frequencies also satisfy the pre-specified error tolerance. The Newton-Raphson iteration results for cracks are shown in Figure 5. 16 (b), where the Newton-Raphson iteration converges after only 6 iteration steps as in Example 5. The final meshes of the 4th order for different Newton-Raphson iteration steps are shown in Figure 5. 17, where the domain near the cracks needs more elements in the 1st, 3rd, and 7th iteration steps.

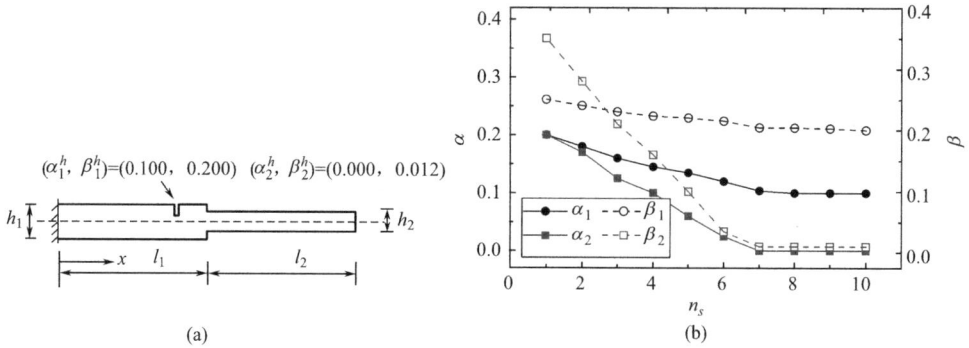

Figure 5. 16　Model and damage detection results for Example 6. Notes: (a) Stepped cantilever beam with a single crack. (b) Newton-Raphson iteration results for cracks

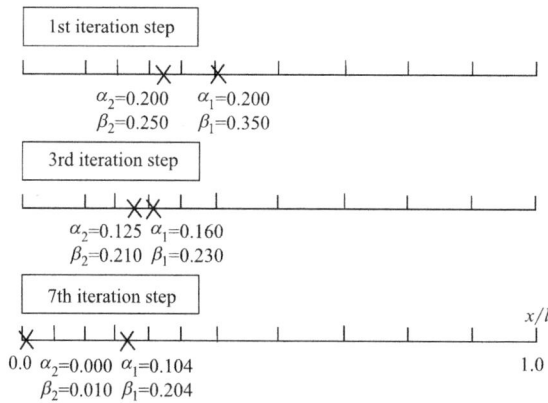

Figure 5. 17　Final meshes of the 4th order computed results for different Newton-Raphson iteration steps of Example 6 (symbol "×" represents the crack)

5. 7. 7　Example 7: Cantilever beam with double cracks

Consider the cantilever beam withtwo cracks in Figure. 18 (a) and the material data

are the same as Example 3 as shown in Equation (5.31). This model had been analysed by the torsional spring model method for simulating the cracks (Lee, 2009). Considering different locations and sizes of cracks as case (a) $\alpha_1 = 0.1$, $\beta_1 = 0.2$, $\alpha_2 = 0.1$, $\beta_2 = 0.4$; case (b) $\alpha_1 = 0.1$, $\beta_1 = 0.4$, $\alpha_2 = 0.2$, $\beta_2 = 0.6$; and case (c) $\alpha_1 = 0.1$, $\beta_1 = 0.6$, $\alpha_2 = 0.2$, $\beta_2 = 0.8$, the first four computed frequencies are listed in Table 5.8 and compared with the solutions obtained with the torsional spring model method. The first four frequencies of forward eigenproblems computed by the proposed method, considered as the actual frequencies, are listed in Table 5.8. In Table 5.8, the differences between the computed and actual frequencies satisfy the pre-specified error tolerance. The Newton-Raphson iteration results of 6 iteration steps for cracks are shown in Figure 5.18 (b), which demonstrates that the proposed method yields a higher convergence rate compared to the conventional FE analysis[19].

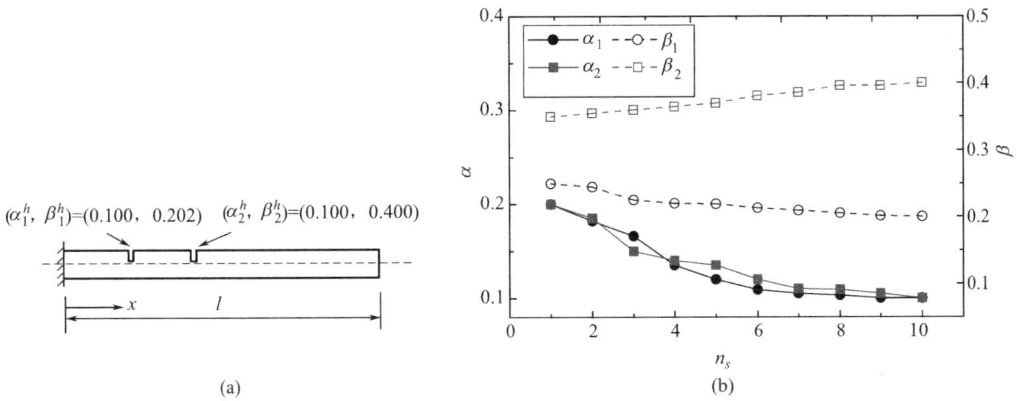

Figure 5.18   Model and damage detection results for Example 7. Notes (a) Cantilever beam with double cracks. (b) Newton-Raphson iteration results for cracks

Computed results for cracks and frequencies of Example 5   Table 5.6

Cracks	Present method		Actual cracks	k	Present method		ω_k ($Tol = 10^{-3}$)
	Computed cracks	ε_α and ε_β			ω_k^h	ε_ω	
α_1	0.00026	0.26E$-$3	0.000	1	21.42747	8.31E$-$04	21.446122
β_1	0.15224	—	—	2	56.80232	4.94E$-$05	56.805177
α_2	0.00059	0.59E$-$3	0.000	3	124.21091	6.54E$-$05	124.219103
β_2	0.65163	—	—	4	196.15611	5.81E$-$04	196.270778

Computed results for cracks and frequencies of Example 6   Table 5.7

Cracks		Present method		Actual cracks	k	Present method		ω_k ($Tol = 10^{-3}$)
		Computed cracks	ε_α and ε_β			ω_k^h	ε_ω	
Case (a)	α_1	0.10061	5.55E$-$04	0.100	1	457.334903	5.80E$-$05	457.308321
	β_1	0.20044	3.67E$-$04	0.200	2	2376.781991	4.30E$-$05	2376.679751
	α_2	0.00059	5.90E$-$04	0.000	3	6649.372784	2.40E$-$05	6649.213179
	β_2	0.01214	—	—	4	12797.592587	5.30E$-$04	12790.812926

continued

Cracks		Present method		Actual cracks	k	Present method		ω_k ($Tol = 10^{-3}$)
		Computed cracks	ε_α and ε_β			ω_k^h	ε_ω	
Case (b)	α_1	0.19931	5.75E−04	0.200	1	457.318404	2.20E−05	457.308321
	β_1	0.20023	1.92E−04	0.200	2	2376.708283	1.20E−05	2376.679751
	α_2	0.00015	1.50E−04	0.000	3	6651.540754	3.50E−04	6649.213179
	β_2	0.98124	—	—	4	12791.746728	7.30E−05	12790.812926
Case (c)	α_1	0.10085	7.73E−04	0.100	1	1344.073836	5.10E−06	1344.066976
	β_1	0.20046	3.83E−04	0.200	2	8300.430701	1.40E−05	8300.314483
	α_2	0.00049	4.90E−04	0.000	3	16335.391083	1.30E−04	16333.267628
	β_2	0.35243	—	—	4	24217.783081	3.60E−05	24216.911236

Computed results for cracks and frequencies of Example 7   **Table 5.8**

Cracks		Present method		Actual cracks	k	Present method		ω_k ($Tol = 10^{-3}$)
		Computed cracks	ε_α and ε_β			ω_k^h	ε_ω	
Case (a)	α_1	0.10026	2.36E−04	0.100	1	419.839607	3.10E−04	419.709187
	β_1	0.20085	7.08E−04	0.200	2	2630.433085	6.10E−05	2630.272577
	α_2	0.10032	2.91E−04	0.100	3	7365.170842	4.50E−05	7364.839379
	β_2	0.40015	1.07E−04	0.400	4	14437.341713	3.60E−04	14432.145781
Case (b)	α_1	0.10012	1.09E−04	0.100	1	419.759672	1.20E−04	419.709187
	β_1	0.39956	3.14E−04	0.400	2	2630.496235	8.50E−05	2630.272577
	α_2	0.20010	8.33E−05	0.200	3	7365.119281	3.80E−05	7364.839379
	β_2	0.60015	9.37E−05	0.600	4	14432.997337	5.90E−05	14432.145781
Case (c)	α_1	0.10045	4.09E−04	0.100	1	419.710617	3.40E−06	419.709187
	β_1	0.60082	5.13E−04	0.600	2	2630.304152	1.20E−05	2630.272577
	α_2	0.19971	2.42E−04	0.200	3	7365.111915	3.70E−05	7364.839379
	β_2	0.80036	2.00E−04	0.800	4	14444.846949	8.80E−04	14432.145781

5.7.8 Example 8: Cantilever beam with triple cracks

Consider the cantilever beam with three cracks in Figure 5.19 (a) with the same material data as Example 7. Three cases are considered for different locations and sizes of cracks as case (a) $\alpha_1 = 0.1$, $\beta_1 = 0.2$, $\alpha_2 = 0.1$, $\beta_2 = 0.4$, $\alpha_3 = 0.1$, $\beta_3 = 0.6$; case (b) $\alpha_1 = 0.1$, $\beta_1 = 0.2$, $\alpha_2 = 0.1$, $\beta_2 = 0.4$, $\alpha_3 = 0.1$, $\beta_3 = 0.8$; and case (c) $\alpha_1 = 0.3$, $\beta_1 = 0.2$, $\alpha_2 = 0.3$, $\beta_2 = 0.4$, $\alpha_3 = 0.3$, $\beta_3 = 0.6$. The first six frequencies of forward eigenproblems computed by the proposed method are listed in Table 5.9 and compared with the solutions obtained using the torsional spring model method. In Table 5.9, the differences between the computed and actual frequencies satisfy the pre-specified error tolerance. The Newton-Raphson iteration results for cracks are shown in Figure 5.19 (b), where the Newton-Raphson iteration converges after 6 iteration steps as in the above three examples, demonstrating that the proposed method yields a higher convergence rate compared to the conventional FE analysis[19]. The final meshes of the 6th order for different Newton-Raphson iteration steps are shown in Figure 5. 20, where the domain near the cracks needs more elements in the 1st, 3rd, and 7th iteration steps. Furthermore, the adaptive FE procedure makes mesh refinement possible.

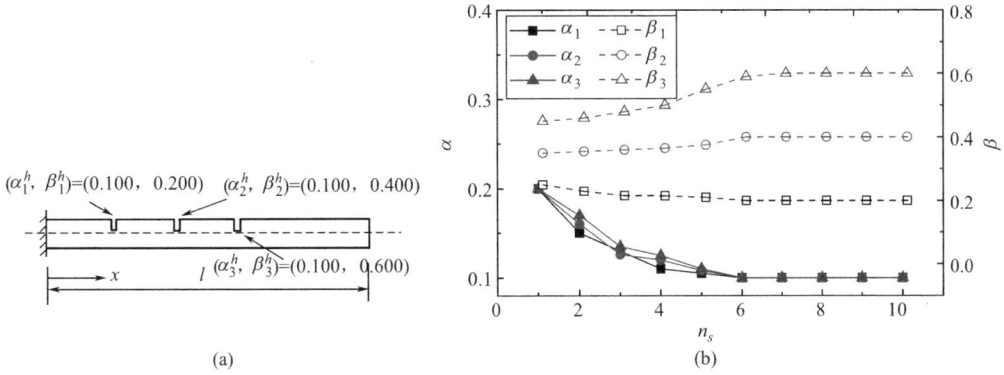

(a)   (b)

Figure 5. 19   Model and damage detection results for Example 8. Notes: (a) Cantilever beam triple cracks. (b) Newton-Raphson iteration results for cracks

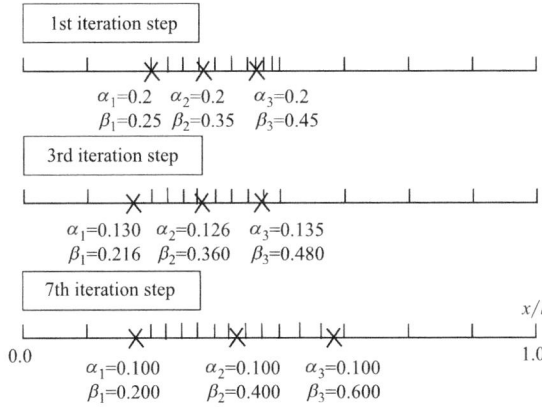

Figure 5. 20   Final meshes of the 6th order computed results for different Newton-Raphson iteration steps of Example 8 (symbol "×" represents the crack)

Computed results for cracks and frequencies of Example 8   Table 5. 9

Cracks		Present method		Actual cracks	k	Present method		ω_k ($Tol = 10^{-3}$)
		Computed cracks	ε_α and ε_β			ω_k^h	ε_ω	
Case (a)	α_1	0. 09952	4. 36E−04	0. 100	1	419. 720125	0. 26E−4	419. 709187
	β_1	0. 20036	3. 00E−04	0. 200	2	2630. 301521	0. 11E−4	2630. 272577
	α_2	0. 10051	4. 64E−04	0. 100	3	7366. 312547	0. 20E−3	7364. 839379
	β_2	0. 39926	5. 29E−04	0. 400	4	14423. 899685	0. 36E−4	14423. 380407
	α_3	0. 10082	7. 45E−04	0. 100	5	23839. 791835	0. 68E−3	23823. 591113
	β_3	0. 60012	7. 50E−05	0. 600	6	35636. 144938	0. 45E−4	35634. 541339

97

continued

Cracks		Present method		Actual cracks	k	Present method		ω_k ($Tol = 10^{-3}$)
		Computed cracks	ε_α and ε_β			ω_k^h	ε_ω	
Case (b)	α_1	0.10064	5.82E−04	0.100	1	419.715498	0.15E−4	419.709187
	β_1	0.20076	6.33E−04	0.200	2	2630.314677	0.16E−4	2630.272577
	α_2	0.09994	5.45E−05	0.100	3	7371.100342	0.85E−3	7364.839379
	β_2	0.40026	1.86E−04	0.400	4	14415.432732	0.22E−4	14415.115577
	α_3	0.10014	1.27E−04	0.100	5	23832.713000	0.36E−3	23824.135951
	β_3	0.79812	1.04E−03	0.800	6	35631.843049	0.42E−4	35630.346532
Case (c)	α_1	0.30026	2.00E−04	0.300	1	419.815499	0.15E−4	419.809187
	β_1	0.20014	1.17E−04	0.200	2	2630.506480	0.75E−4	2630.309132
	α_2	0.29881	9.15E−04	0.300	3	7369.768996	0.71E−3	7364.539463
	β_2	0.40011	7.86E−05	0.400	4	14432.355632	0.13E−4	14432.168001
	α_3	0.30023	1.77E−04	0.300	5	23660.686382	0.14E−3	23657.374210
	β_3	0.60074	4.63E−04	0.600	6	35641.050974	0.62E−4	35638.841304

5.8 Conclusions

The conclusions can be summarised as follows:

(1) In this study, a new adaptive FEM methodology was presented for accurate computation of both the frequencies and modes of cracked Euler-Bernoulli beams, and the adaptive analysis technology has been developed and applicated for the reliable computation of the locations and sizes of multiple cracks. Some key techniques are utilized, i. e. , adaptive FE analysis for eigensolutions, Newton-Raphson iteration, and damage refinement techniques, based on the conventional frequency measurement method for damage detection, which has yielded a simple and practical adaptive FE procedure that finds sufficiently fine meshes for the accurate locations and sizes of multiple cracks to match the pre-specified error tolerance.

(2) Numerical examples are provided, including ones known to be representative of a non-uniform and geometrically stepped Euler-Bernoulli beam with multiple cracks, to demonstrate the accuracy, reliability, and effectiveness of the proposed adaptive FE algorithm and procedure. Based on frequency measurements for damage detection, the inverse eigenproblem computation makes full use of the forward eigenproblem computation for frequency solutions. As a result, making the two forward and inverse complementary parts of the research series work together, the proposed FE procedure reduces the cost of computation and improves the accuracy of the solutions for determining the locations and sizes of cracks in beams.

The present chapter is limited to Euler-Bernoulli beam beams with cracks, but with conventional numerical treatments of integration of beams, the present method can also solve some frame structure problems in an indirect way. Looking forward, a very welcoming and encouraging feature of this presented methodology is that it can readily be extended to damage detection problems of frame structure with multiple cracks as engineering practice, which will be addressed in future study.

References

[1] Wang Z, Lin R M, Lim M K, et al. Structural damage detection using measured FRF data [J]. Computer Methods in Applied Mechanics and Engineering, 1997, 147 (1-2): 187-197.

[2] Hassiotis S, Jeong G D. Assessment of structural damage from natural frequency measurements [J]. Computers and Structures, 1993, 49 (4): 679-691.

[3] Pawar P M, Ganguli R. Genetic fuzzy system for damage detection in beams and helicopter rotor blades [J]. Computer Methods in Applied Mechanics and Engineering, 2003, 192 (16-18): 2031-2057.

[4] Zacharias J, Hartmann C, DelgadoA, et al. Damage detection on crates of beverages by artificial neural networks trained with finite-element data [J]. Computer Methods in Applied Mechanics and Engineering, 2004, 193 (6-8): 561-574.

[5] Yan W, Chen W Q, Cai, et al. Quantitative structural damage detection using high-frequency piezoelectric signatures via the reverberation matrix method [J]. International Journal for Numerical Methods in Engineering, 2007, 71 (5): 505-528.

[6] Farrar C R, Doebling S W, Nix D A, et al. Vibration-based structural damage identification [J]. Philosophical Transactions of the Royal Society, 2001, 359 (1778): 131-149.

[7] Wang D, Liu W, Zhang H, et al. Superconvergent isogeometric free vibration analysis of Euler-Bernoulli beams and Kirchhoff plates with new higher order mass matrices [J]. Computer Methods in Applied Mechanics and Engineering, 2015, 286: 230-267.

[8] Kaveh A, Dadfar B. Eigensolution for free vibration of planar frames by weighted graph symmetry [J]. International Journal for Numerical Methods in Engineering, 2007, 69 (6): 1305-1330.

[9] Litewka P, Rakowski J. Free vibrations of shear-flexible and compressible arches by FEM [J]. International Journal for Numerical Methods in Engineering, 2001, 52 (3): 273-286.

[10] Labib A, Kennedy D, Featherston C, et al. Free vibration analysis of beams and frames with multiple cracks for damage detection [J]. Journal of Sound and Vibration, 2014, 333 (20): 4991-5003.

[11] Nandwana B P, Maiti S K. Detection of the location and size of a crack in stepped cantilever beams based on measurements of natural frequencies [J]. Journal of Sound and Vibration, 1997, 203 (3): 435-446.

[12] Chaudhari T D, Maiti S K. A study of vibration of geometrically segmented beams with and without crack [J]. International Journal of Solids and Structures, 2000, 37 (5): 761-779.

[13] Greenberg L, Marletta M. Algorithm 775: The code SLEUTH for solving fourth order Sturm-Liouville problems [J]. ACM Transactions on Mathematical Software, 1997, 23 (4): 453-493.

[14] Caddemi S, Morassi A. Multi-cracked Euler-Bernoulli beams: mathematical modeling and exact solutions [J]. International Journal of Solids and Structures, 2013, 50 (6): 944-956.

[15] Caddemi S, Caliò I. Exact reconstruction of multiple concentrated damages on beams [J]. Acta Mechanica, 2014, 225 (11): 3137-3156.

[16] Hsu M H. Vibration analysis of edge-cracked beam on elastic foundation with axial loading using the differential quadrature method [J]. Computer Methods in Applied Mechanics and Engineering, 2005, 194 (1): 1-17.

[17] Rizos P F, Aspragathos N, Dimarogonas A D, et al. Identification of crack location and magnitude in a cantilever beam from the vibration modes [J]. Journal of Sound and Vibration, 1990, 138 (3): 381-388.

[18] Chinchalkar S. Determination of crack location in beams using natural frequencies [J]. Journal of Sound and Vibration, 2001, 247 (3): 417-429.

[19] Lee J. Identification of multiple cracks in a beam using natural frequencies [J]. Journal of Sound and Vibration, 2009, 320 (3): 482-490.

[20] Zienkiewicz O C, Zhu J. The superconvergent patch recovery and a posteriori error estimates. Part 1: The recovery technique [J]. International Journal for Numerical Methods in Engineering, 1992, 33 (7): 1331-1364.

[21] Zienkiewicz O C, Zhu J. The superconvergent patch recovery and a posteriori error estimates. Part 2: Error estimates and adaptivity [J]. International Journal for Numerical Methods in Engineering, 1992, 33 (7): 1365-1382.

[22] Wiberg N E, Bausys R, Hager P, et al. Adaptive h-version eigenfrequency analysis [J]. Computers and Structures, 1999, 71 (5): 565-584.

[23] Wiberg N E, Bausys R, Hager P, et al. Improved eigenfrequencies and eigenmodes in free vibration analysis [J]. Computers and Structures, 1999, 73 (1-5): 79-89.

[24] Wang Y, Liu Z, Yang H, et al. Finite element analysis for wellbore stability of transversely isotropic rock with hydraulic-mechanical-damage coupling [J]. Science China Technological Sciences, 2017, 60 (1): 133-145.

[25] Wang Y, Zhuang Z, Liu Z, et al. Finite element analysis for inclined wellbore stability of transversely isotropic rock with HMCD coupling based on weak plane strength criterion [J]. Science China Technological Sciences, 2017, 10. 1007/s11431-016- 0460-2. .

[26] Yuan S, Wang Y, Ye K, et al. An adaptive FEM for buckling analysis of non-uniform Bernoulli-Euler members via the element energy projection technique [J]. Mathematical Problems in Engineering, 2013, 10. 1155/2013/461832. .

[27] Yuan S, Wang Y, Xu J, et al. New progress in self-adaptive FEMOL analysis of 2D free vibration problems [J]. Enginering Mechanics, 2014, 31 (1): 15-22.

[28] Dimarogonas A D. Vibration of cracked structures: a state of the art review [J]. Engineering Fracture Mechanics, 1996, 55 (5): 831-857.

[29] Moezi S A, Zakeri E, Zare A, et al. On the application of modified cuckoo optimization algorithm to the crack detection problem of cantilever Euler-Bernoulli beam [J]. Computers and Structures, 2015, 157: 42-50.

[30] Guan H, Karbhari K M. Improved damage detection method based on element modal strain damage index using sparse measurement [J]. Journal of Sound and Vibration, 2008, 309 (3): 465-494.

[31] Owolabi G M, Swamidas A S J, Seshadri R, et al. Crack detection in beams using changes in frequencies and amplitudes of frequency response functions [J]. Journal of Sound and Vibration, 2003, 265 (1): 1-22.

[32] Maghsoodi A, Ghadami A, Mirdamadi H R, et al. Multiple-crack damage detection in multi-step beams by a novel local flexibility-based damage index [J]. Journal of Sound and Vibration, 2013, 332

(2): 294-305.

[33] Al-Said S M. Crack identification in a stepped beam carrying a rigid disk [J]. Journal of Sound and Vibration, 2007, 300 (3): 863-876.

[34] Al-Said S M. Crack detection in stepped beam carrying slowly moving mass [J]. Journal of Sound and Vibration, 2008, 14 (12): 1903-1920.

[35] Morassi A. Identification of a crack in a rod based on changes in a pair of natural frequencies [J]. Journal of Sound and Vibration, 2001, 242 (4): 577-596.

[36] Lele S P, Maiti S K. Modelling of transverse vibration of short beams for cracks detection and measurement of crack extension [J]. Journal of Sound and Vibration, 2002, 257 (3): 559-583.

[37] Nikolakopoulos P G, Katsareas D E, Papadopoulos C A, et al. Crack identification in frame structures [J]. Computers and Structures, 1997, 64 (1-4): 389-406.

[38] Narkis Y. Identification of crack location in vibrating simply supported beams [J]. Journal of Sound and Vibration, 1994, 172 (4): 549-558.

[39] Dado M H. A comprehensive crack identification algorithm for beam under different end conditions [J]. Applied Acoustics, 1997, 51 (4): 381-398.

[40] Hu J, Liang R Y. An integrated approach to detection of cracks using vibration characteristics [J]. Journal of the Franklin Institute, 1993, 330 (5): 841-853.

[41] Patil D P, Maiti S K. Detection of multiple cracks using frequency measurements [J]. Engineering Fracture Mechanics, 2003, 70 (12): 1553-1572.

[42] Ruotolo R, Surace C. Damage assessment of multiple cracked beams: numerical results and experimental validation [J]. Journal of Sound and Vibration, 1997, 206 (4): 567-588.

[43] Shifrin E I, Ruotolo R. Natural frequencies of a beam with an arbitrary number of cracks [J]. Journal of Sound and Vibration, 1999, 222 (3): 409-423.

[44] Labib A, Kennedy D, Featherston C A, et al. Crack localisation in frames using natural frequency degradations [J]. Computers and Structures, 2015, 157: 51-59.

[45] Yuan S, Ye K, Wang, et al. Adaptive finite element method for eigensolutions of regular second and fourth order Sturm-Liouville problems via the element energy [J]. Engineering Computations (Accepted), 2017.

[46] Zienkiewicz O C, Taylor R L. The Finite Element Method [M]. fifth edition. Oxford: Butterworth-Heinemann, 2000.

[47] Clough R W, Penzien J. Dynamics of Structures [M]. second edition. New York: McGraw-Hill, 1993.

Chapter 6
Adaptive finite element method for
stability disturbance of cracked beams

6. 1   Introduction

Damage to the skeletal structure of a building is a significant problem in practical engineering, and primarily occurs in the form of cracks[1, 2]. Crack damage can change the mechanical properties of the entire structure and affect the safety and applicability of the structure[3]. Studying the dynamic characteristics of a beam structure with multiple cracks and accurately predicting the buckling bearing capacity can effectively guarantee the safety of a structure over its entire life cycle[4]. Curved beams are widely employed in space, mechanical, and civil engineering[5]. Owing to their complex geometry, curved beams are prone to elastic buckling instability[6]. Accurately evaluating the buckling load of various circularly curved beam lines and different curved beam angles for deep or shallow beams is an important aspect of structural disaster analysis. The existence of crack damage in curved beams increases the difficulty of accurately predicting the buckling instability and bearing capacity, and theoretical models and analytical methods are often simplified and cannot be used to perform effective analyses[7-9]. Therefore, an effective solution is required to overcome these problems.

The accurate prediction of the buckling load considering different crack locations, magnitudes, and numbers, and the analysis of the influence of crack damage on buckling instability[10, 11] have received significant attention in theoretical research and engineering practice. Considering the effect of cracks on the elastic buckling of circularly curved beams under different working conditions, Kim et al. [12] introduced a precise displacement field to study the influence of the thickness curvature and bending condition on the free vibration of the curved beam. The free vibration of circularly curved beams with variable cross-sections was analysed using the spectral element method, considering the effects of various boundary conditions on the frequency parameters[13]. By establishing a curved beam element with cracks, finite element analyses were used to analyse the in-plane dynamic response of isotropic circularly curved beams with cracks under moving loads, and to study the influence of the crack location on beam dynamics[14]. The effects of a unilateral crack and its location on the load, natural frequency, and dynamic stability of circularly curved beams were studied using the energy method[15]. The effects of crack location and depth on the stability of a system were also analysed using the finite element and multi-scale methods[16]. The vibration behaviour of a simply supported beam with cracks in its middle span was analysed to determine the effect of crack length on the mode shape and amplitude[17]. In general, the natural frequency of a beam decreases with the increase in crack depth and is related to the crack location in the material[18]. Furthermore, modal analysis revealed that the size of a crack has a significant influence on the dynamic characteristics of the free end of a beam[19]. Some researchers simulated the in homogeneity of the material using the equivalent intrinsic strain and used Eringen's nonlocal elastic model to determine that the crack size has a significant influence on the brittle fracture characteristics of a material[20, 21]. Others established the free vibration model of a beam with cracks and determined that the position and depth of a crack and the crack size have different effects on the free vibration frequency[22, 23]. In addition, the geometric free vibration of an Euler-Bernoulli curved beam was analysed, and a theoretical analysis of its frequency accuracy was performed[24]. However, despite extensive research, it is still difficult to effectively apply these findings under complex working conditions such as in beams with variable curve types, multi crack damage, diverse boundary conditions, etc.

High-precision solutions of eigenvalues and eigenfunctionsare of considerable importance in research and engineering. For example, the representative dynamic analysis of structures based on high-precision solutions of the natural frequency (eigenvalue) or vibration mode (eigenfunction)[25] were used to accurately identify the number, size, and location of damages in cracked structures[26, 27]. Various analytical and theoretical models have been developed to obtain high-precision free vibration solutions of beam members. For example, the free vibration of cylindrical shells under homogeneous boundary conditions was studied[28] based on Flugge elastic thin shell theory[29]. After introducing the general form of a displacement function, the dispersion diagram of the frequency and wave number can be obtained using the displacement control equation, which can be transformed into a zero-point solution of an eighth-order matrix determinant[30] by combining the boundary conditions. An exact analytical solution can be obtained by deducing the element stiffness matrix and mass matrix of a horizontal curved beam[31]. Considering the effects of the moment of inertia and shear deformation, the spectral element method was proposed to study the free vibration of circularly curved beams[32]. A formula for the free vibration analysis of a curved beam in functional gradient space was presented based on the first-order shear deformation theory[33]. As the dynamic response of any structure includes its inherent characteristics, the inherent frequency of a beam with a breathing crack can be determined based on the relationship between frequency and displacement. The transverse vibration of a simply supported beam with a breathing crack was predicted based on the vibration theory of a continuous beam with breathing cracks[34]. A simple finite element model was used to simulate the experimental vibration behaviour of a free-free beam with a breathing crack considering a sinusoidal input force[35]. The adaptive generalised finite element method has also been applied to the longitudinal free vibration analysis of straight bars and trusses[36]. Nevertheless, owing to the disturbance effect of each mode of crack damage formation, a more accurate solution is required for analysis.

Numerical computation is a reasonable choice for analysing the dynamic performance of complex structures. The finite element method has been widely used to determine the elastic buckling load and buckling mode of curved beams with cracks[15, 37, 38]. However, the accuracy of the solution depends on the quality of mesh refinement, which inevitably results in errors[39]. Adaptive mesh refinement can effectively optimise the mesh distribution, and can be employed for solving problems such as the elastic buckling of straight beams[40], vibration of damaged beams[41], vector Sturm-Liouville problems and free vibration of curved beams[42], hp-version adaptive finite element algorithm for eigensolutions[43], and damage and fracture in rock with coupling of multiphysical fields[44]. A general finite element formula was developed considering the non-classical and non-uniform torsion of beams composed of anisotropic materials[45]. The natural frequencies of numerous hollow beam sizes can be computed using the Euler-Bernoulli, Rayleigh, and Timoshenko beam theories. A complete derivation of the elastic solution and beam theory is also available in the literature[46]. A four-node curved shell finite element method was proposed for the geometric nonlinear analysis of plane curved beams[47]. A finite element formula for a non-uniform shear warping model of thin-walled beams was established to determine the precise distribution of normal stress and shear stress on the beam section, including the shear lag phenomena[48]. Owing to the complexity of the interpolation function assumption and classical finite element analysis, an effective torsion element was established to describe the constrained torsion of thin-walled beams based on the first-order torsion theory[49]. Compared with the conventional finite element method, the adaptive finite element method can provide high-precision solutions within the error limits specified by the user[50]. Adaptive algorithms can be used to optimise the mesh and improve solution accuracy[51]; some examples of adaptive algorithms include the p-version adaptive method, which can be used to improve the element order[52], the h-version adaptive method, which can be used to increase the mesh density[5], and the ph-version adaptive method[53].

In this study, we propose an h-version finite element mesh adaptive analysis method for a variable section Euler-Bernoulli beam to solve the elastic buckling problem of a circularly curved beam with crack damage. In addition, we establish a cross-section damage defect comparison scheme for circularly curved beams to simulate the magnitude (depth), location, and number of cracks. The h-version adaptive finite element method[41] for non-uniform Euler-Bernoulli beams is used to solve the crack damage problem of a circularly cracked beam under elastic buckling. Using the proposed method, it is possible to obtain final optimised meshes and high-precision buckling loads and modes that satisfy the pre-specified error tolerance. Several numerical examples are illustrated to solve the elastic buckling problem of circularly curved beams. The adaptive mesh refinement and convergence of the elastic buckling solution are discussed as well. The influence of the elastic buckling load and buckling mode of circularly curved beams on the subtended angle, damage location, and number and magnitude of cracks is analysed to verify the efficacy of the proposed adaptive mesh refinement method.

The rest of this chapter is organised as follows. The elastic buckling problem of circularly curved beams based on Euler-Bernoulli beam theory is introduced in Section 6.2. The characterisation method of micro-crack damage in circularly curved beams is presented in Section 6. 3. The local mesh refinement techniques and procedure are introduced in Section 6. 4. Numerous representative numerical examples are presented in Section 6. 5 to demonstrate the performance of the proposed method. Finally, the main conclusions of the study are summarised in Section 6. 6.

6. 2   Elastic buckling of circularly curved beams based on Euler-Bernoulli theory

Consider the curved beam shown in Figure 6. 1: the neutral axis coordinate of the curved beams is s; the coordinate system is x-y; the tangential lies along the x axis, the normal lies along the y axis, and the in-plane displacement w occurs along the y axis; the angular coordinates are for θ. The radius of curvature of the curved beam is $R(s)$, $I(s)$ is the inertial moment, l is the length of the beam, h is the height of the beam, b is the thickness of the beam, E is the elastic modulus of the material, h_c is the crack depth, and δ_c is the width of the crack damage across the cross-section. These parameters extend through the cross-section of the beam along the thickness direction.

This study discusses the elastic buckling problem of a curved beam. Figure 6. 2 shows the cross-section damage and loading diagram of a circularly curved beam with crack damage, where in θ_0 represents the subtended angle and q represents the even load it is subjected to. The relationship of the curved beam between the neutral axis coordinate s and the angular coordinate θ is:

$$s = \theta R, \ 0 < \theta < \theta_0 \tag{6.1}$$

where, the constant R is the radius of the circularly curved beam, and the crack damage occurs at the angular coordinate θ_c.

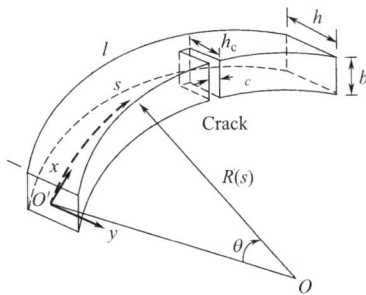

Figure 6. 1   Coordinate systems and symbols of cracked curved beam

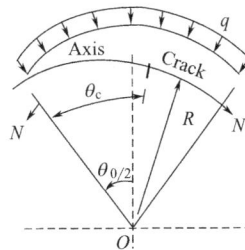

Figure 6. 2   Diagram of cross-section damage defect and loading for circularly curved beam with crack damage

The circularly curved beam can be classifiedbased on the standard shown in Table 6. 1[15, 54]. As shown, curved beams can be divided into thick, medium thick, and thin beams based on the radius-thickness ratio R/h. They can also be divided into shallow, medium deep, deep, and ultra-deep beams based on the subtended angle θ_0. The height of a thin curved beam is relatively low, and it is extremely prone to elastic buckling instability owing to crack damage. The Euler-Bernoulli beam model is suitable for thin beams, ignoring shear deformation[55]. Therefore, the elastic buckling problem of a thin curved beam is analysed herein based on the Euler-Bernoulli beam model.

Categories of circularly curved beams based on radius-thickness ratio and subtended angle

Table 6. 1

R/h	Thick	Medium thick		Thin
	$R/h < 40$	$R/h = 40$		$R/h > 40$
θ_0	Shallow	Medium deep	Deep	Ultra-deep
	$\theta_0 < 40°$	$\theta_0 = 40°$	$40° < \theta_0 \leqslant 180°$	$\theta_0 > 180°$

Herein, the differential governing equation used for the elastic buckling of a curved beam is[40]:

$$Lv \equiv (EI(s)w''(s))'' = -\lambda(P(s)w'(s))', \ 0 < s < l \qquad (6.2)$$

where, L is the corresponding differential operator, $()' = d()/ds$, $P(s)$ is the load function, λ is the ultimate buckling load factor, $\lambda P(s)$ is the ultimate buckling load q_c, and $w(s)$ is the buckling mode. Generally, the simply supported boundary conditions of a curved beam are:

$$w(0) = 0, \quad w(l) = 0 \qquad (6.3)$$

Solving the eigenvalue in Equation (6.2) under the given finite element mesh π, the conventional finite element will set up the following line matrix eigenvalue equation:

$$\boldsymbol{KD} = \lambda \boldsymbol{K}_g \boldsymbol{D} \qquad (6.4)$$

where $\boldsymbol{D}$ is the finite element solution of the buckling mode, and $\boldsymbol{K}$ and $\boldsymbol{K}_g$ are the static stiffness matrix and geometric stiffness matrix, respectively. The finite element solution (q_c^h, w^h) under the current mesh can be obtained based on the inverse power iteration method[56].

6.3　Characterisation method for micro-crack damage in circularly curved beams

Owing to crack damage in a curved beam, its cross-section weakens and its flexural rigidity is attenuated. The comparison method of crack damage cross-section defect[41] is adopted herein; the cross-sectional stiffness at the crack is:

$$EI_c = \frac{Ebh^3(1-\alpha)^3}{12} \qquad (6.5)$$

where $\alpha = h_c/h$ is the crack damage rate of the cross-section of the beam; $\alpha = 0$ indicates that the cross-section of the beam is complete and has no damage. As the section damage width δ_c of a micro-crack is set to a smaller value, the following setting is adopted herein.

$$\delta_c = 0.01 \times Tol \qquad (6.6)$$

where *Tol* is the pre-specified error tolerance of the elastic buckling solution.

6. 4   Local mesh refinement techniques and procedure

In finite elementcomputations, there are existing superconvergence points with a higher convergence order than the solution for a given mesh[56]. By combining these superconvergence points with element patch and the interpolation technology of high-order form functions, the accuracy of the finite element solution can be improved and a global superconvergence solution can be obtained[41, 57, 58]. Herein, the finite element solution of the buckling mode (displacement) with the current mesh is obtained for the elastic buckling problem of the circularly curved beam, and the superconvergent patch recovery method in finite element post-processing is used to obtain the superconvergent solution of the buckling mode.

$$w^*(x) = \boldsymbol{Pa} \tag{6.7}$$

where $\boldsymbol{P}$ is the given function vector and $\boldsymbol{a}$ is the undetermined coefficient vector.

The buckling load can be obtained using modal solutions and Rayleigh quotient computation[59]. The superconvergent solution of the buckling modes can be used to obtain the superconvergent solution of the buckling load q_c^*.

$$q_c^* = \sqrt{\frac{a(w^*, w^*)}{b(w^*, w^*)}} \tag{6.8}$$

where $a()$ and $b()$ are the strain energy and kinetic energy inner product, respectively[41]. The superconvergent solution of the buckling load computed by the Rayleigh quotient has a higher convergence order than the superconvergent solution of the buckling modes[59]. Therefore, we perform error estimation and control for the buckling modes to obtain high-precision buckling mode solutions, ensuring their accuracy.

The error estimation of the finite element solution of the buckling mode under the current mesh in the form of the energy mode can be carried out by introducing the super-convergent solution of the buckling mode[41, 56]:

$$\| e^* \| \leqslant Tol \cdot [(\| w^h \|^2 + \| e^* \|^2)/n_e]^{1/2} \tag{6.9}$$

where n_e is the number of patch elements, $e^* = w^* - w^h$, $\| e^* \|$ is the energy norm. In Equation (6.9), the error estimate can be expressed in the following relative error form:

$$\xi = \frac{\| e^* \|}{\overline{e}} < 1, \ \overline{e} = Tol \cdot [(\| w^h \|^2 + \| e^* \|^2)/n_e]^{1/2} \tag{6.10}$$

where ξ is the relative error value.

The mesh can be optimised to reduce and control the buckling mode error and achieve the pre-specified solution accuracy by using the error estimation of the buckling mode. The method employed herein is to judge the vibration mode error of each finite element e. If the error does not satisfy Equation (6.10), the error of the buckling mode solution on the element is too high and mesh optimisation is required. In this study, we employ the

109

uniform mesh by subdividing the element to increase the degree of freedom of the model and reduce the error of the element solution[41]. The new element length generated by the current subdivision is related to the current error and element order; therefore, the new element length can be estimated using the current error:

$$h_{new} = \xi^{-1/m} h_{old} \tag{6.11}$$

where h_{old} is the length of the current element e, and h_{new} is the length of the new element.

Using these techniques, the elastic buckling disturbance and local mesh refinement of circularly curved beams with multiple micro-cracks can be analysed using the procedure shown in Figure 6.3. The FE model of the beam with crack damage is established using the characterisation method for micro-crack damage (including the crack size (depth), location, and number), and the FE solution of the bucking load and mode with the current mesh is obtained by solving the model. Subsequently, error estimation and mesh subdivision optimisation of the region around the crack damage are performed considering the buckling modes of each order, and the FE mesh is updated. Finally, the effects of micro-crack damage on the elastic buckling disturbance are analysed using high-precision solutions of beams with crack damage to determine the impact of different types of micro-crack damage (including the crack size (depth), location, and number) on the elastic buckling disturbance.

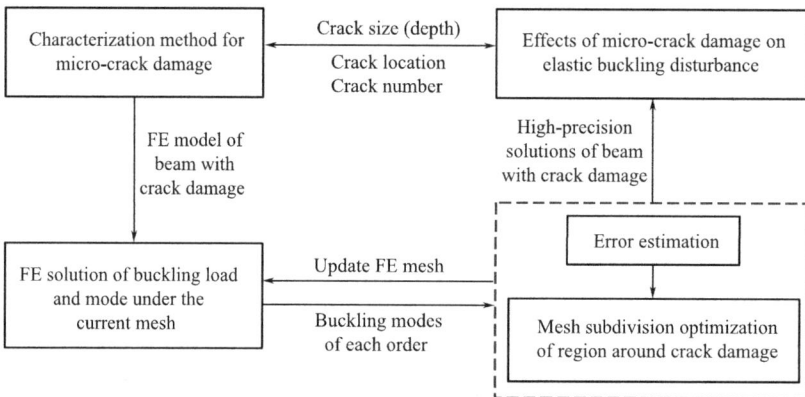

Figure 6.3 Analysis procedure of elastic buckling disturbance and local mesh refinement of circularly curved beams due to multiple micro-crack damage

6.5 Numerical examples

The proposed numerical method was programmed in Fortran 90. In this section, we present several representative numerical examples of elastic buckling in circularly curved beams. The convergence of the adaptive mesh division and elastic buckling solution is discussed. The effects of the subtended angle, damage location, number, and magnitude of cracks in curved beams on the elastic buckling load and buckling mode are analysed, and the effectiveness of the adaptive mesh division is verified. All the numerical examples presented herein used an element with three orders, an initial mesh with two elements, and an initial error limit of $Tol = 10^{-4}$.

6.5.1   Example 1: Verification for eigensolutions and refined meshes of elastic buckling of uncracked curved beam

Consider a circularly curved beamthat is simply supported at both ends, with geometric and physical parameters:

$$\theta = \pi/6,\ R = 0.254\ \text{m},\ h = 0.006\ \text{m},\ b = 0.006\ \text{m},\ E = 68.95\ \text{GPa} \qquad (6.12)$$

The radius-thickness ratio R/h of this beam is 42.33, which indicates that it is a typical thin curved beam.

Herein, the elastic buckling of the curved beam is solved, and the solutions of the buckling load and buckling mode are obtained. In previous studies, the layered composite beam model and finite element method[60] or the analytical method of the theoretical model[9] have been adopted to analyse the elastic buckling of a beam and determine the buckling load value q_{nc} of an uncracked curved beam. The buckling load values obtained using these methods are compared with those obtained using the proposed method in Table 6.2. To analyse the influence of the mesh number on the convergence of the solution, the conventional finite element method is used to solve the numerical model with 4, 5, and 6 elements (sparse uniform meshes), and 25, 50, and 100 elements (dense uniform meshes), and determine the buckling load value. The results indicate that the solution tends to become stable as the number of elements increases, and convergent solutions are obtained closer to 100 elements. Using the adaptive finite element method herein, we obtained the optimised meshes and converged solutions of 16 elements with this mesh. The buckling load solution (64.406 kN/m) coincides well with the analytical solution of the theoretical model (64.966 kN/m). Notably, the solutions obtained using 4, 5, and 6 elements in the composite beam model differ significantly from the analytical solutions owing to the layered model of the beam model.

Convergence for number of elements and buckling loads results   Table 6.2

Numbers of element	q_{nc} (kN/m)			
	Conventional finite element method	Composite beam model[1]	Analytical method [2]	Proposed method
4	64.450	77.566		
5	64.450	77.181	64.966	64.406
6	64.436	77.046		(16 elements)
25	64.440	—		
50	64.408	—		
100	64.406	—		

Source: [1] Results from paper [60]; [2] Results from paper [9].

The buckling mode solution obtained using the adaptive mesh is given in Figure 6.4. For ease of understanding and analysis, the buckling mode results are normalised (the maximum mode value is 1). As the physical properties and geometric forms of the beam are symmetrical, the buckling modes shown in Figure 6.4 (a) are also symmetrical. The distribution of the adaptive mesh (element end nodes) is shown on the horizontal axis. As shown, the distribution of the elements adapting a change of mode also have a left-right symmetry and use relatively dense meshes at both ends of the boundary region. The algorithm automatically optimises the non-uniform mesh, using a sparse mesh in the region where the buckling mode change is gradual and a relatively fine mesh in the region where the buckling mode change is severe. This avoids the redundancy of using a uniform fine mesh across the entire domain. To test the accuracy of the buckling mode solution, the conventional finite element method was employed with 2500 elements (highly dense uniform mesh) to obtain a high-precision analytical solution of the buckling mode w_n. The morphology of the high-precision buckling mode is the same as that in Figure 6.4 (a). Figure 6.4 (b) shows the distribution curve of the difference between the buckling modes determined using the proposed adaptive method and the high-precision analytical solution. The maximum difference is 2.17×10^{-5}, which is less than the pre-specified error tolerance of 10^{-4}, thereby verifying the accuracy of the proposed method.

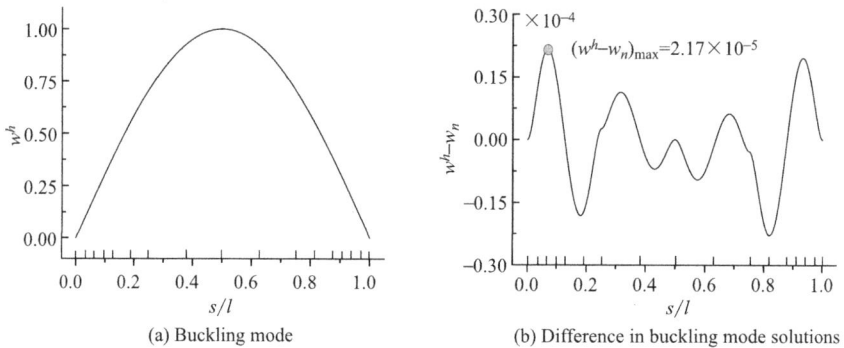

(a) Buckling mode   (b) Difference in buckling mode solutions

Figure 6.4   Buckling mode with adaptive refinement mesh

The differences between the distributions of the high-precision buckling mode solutions and the buckling mode solutions obtained through conventional finite element analysis with 4, 5, 6, 25, 50, and 100 elements are shown in Figure 6.5. The distribution of each mesh is shown on the horizontal coordinate axes in the figures. The maximum difference in each domain with 4, 5, 6, 25, 50, and 100 elements is 1.23×10^{-1}, 1.23×10^{-1}, 4.12×10^{-3}, 7.70×10^{-3}, 1.62×10^{-3}, and 2.63×10^{-6}, respectively. As the density of the mesh increases, the error of the buckling mode solution gradually decreases. A solution that satisfies the pre-specified error tolerance of 10^{-4} is only obtained when enough elements (100 elements) are present. The adaptive method proposed herein adopts a mesh with non-uniform distribution containing 16 elements, thereby avoiding the element redundancy of a high-density uniform distribution mesh and improving the computational efficiency.

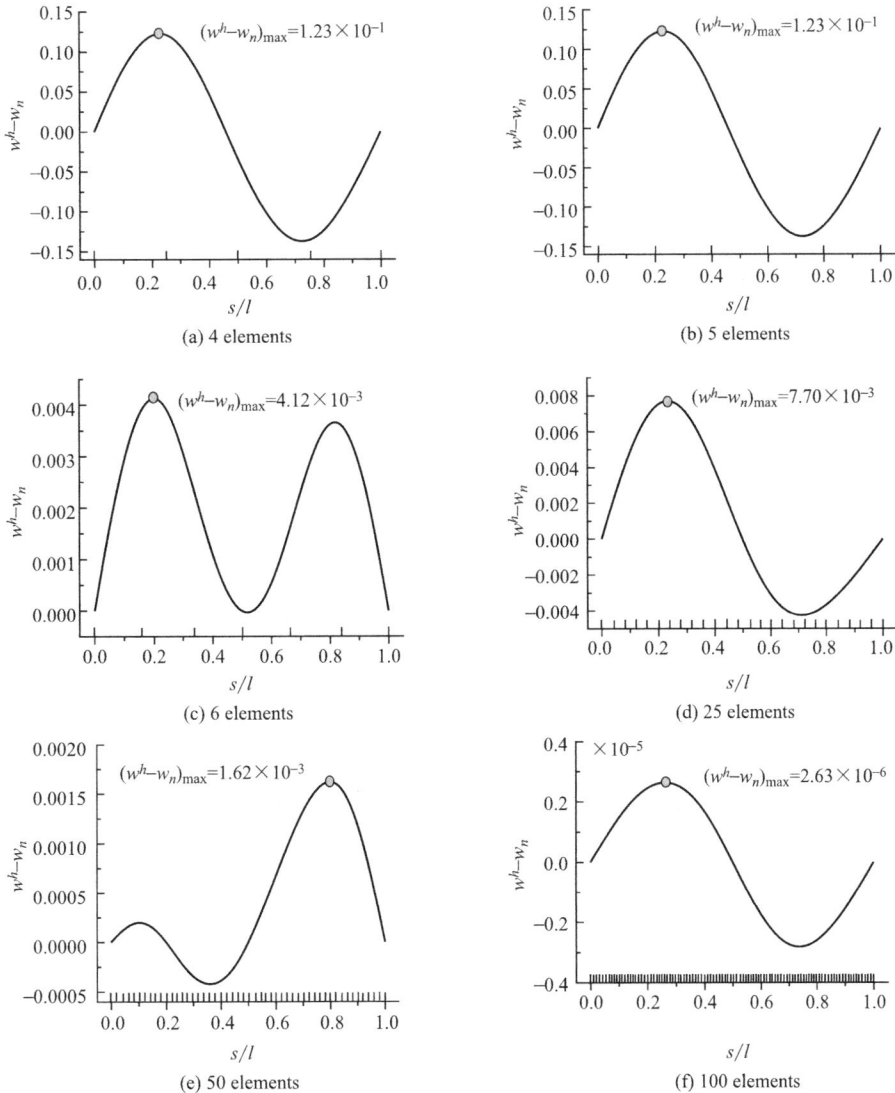

Figure 6.5   Convergence of buckling modes on refined meshes

6.5.2   Example 2: Buckling loads under different subtended angles

To test the applicability of the proposed method in solving the elastic buckling of curved beams with different geometric forms, circularly curved beams with different subtended angles (shallow beams and deep beams) were analysed herein. Consider a circularly curved beam that is simply supported at both ends, with the geometric and physical parameters:

$$R = 0.254 \text{ m}, \ h = 0.005 \text{ m}, \ b = 0.005 \text{ m}, \ E = 68.95 \text{ GPa} \tag{6.13}$$

113

The radius-thickness ratio R/h of this beam is 50.8, which indicates that it is a thin curved beam.

Various approaches have been used to calculate the buckling load of curved beams in previous studies including the energy method[15], finite element method[60], and analytical method of the theoretical model[9]; the buckling load values of curved beams with subtended angles of $\pi/6$, $\pi/5$, $\pi/4$, $\pi/3.5$, and $\pi/2$ were calculated using these methods and the proposed method, and the results are compared in Table 6.3. The buckling load values obtained using the proposed method are in good agreement with those obtained using the analytical method for each subtended angle. This reflects the applicability of the proposed method to the elastic buckling solution of various curved beams with different geometric forms, such as shallow beams and deep beams.

Buckling loads of curved beams with hinged-hinged supports under different subtended angles

Table 6.3

Subtended angle	q_{nc} (kN/m)			
	Energy method[1]	Compositebeam model[2]	Analytical method[3]	Proposed method
$\pi/6$	31.604	36.489	31.331	31.060
$\pi/5$	21.945	24.723	21.690	21.569
$\pi/4.5$	17.773	19.861	17.527	17.471
$\pi/4$	14.040	15.604	13.803	13.804
$\pi/3.5$	10.745	11.906	10.517	10.569
$\pi/2$	3.481	3.854	3.286	3.4511

Source: ① Results from paper [15], ② Results from paper [60], ③ Results from paper [9].

6.5.3   Example 3: Different locations of single crack damage

The elastic buckling solution considering a typical crack damage location and magnitude was solved using the beam model of crack damage and the proposed adaptive method. The influence of crack damage on the solution was analysed as well. Consider a circularly curved beam that is simply supported at both ends; the subtended angle $\theta = \pi$ and the other basic geometrical and physical parameters are the same as those in Equation (6.12).

Considering a damage depth of $h_c = 0.003$ m, $(h_c/h = 0.5)$, the elastic buckling loads when the crack damage is at the left half span of a curved beam with circular angle coordinates θ_c of 0, $\pi/12$, $\pi/6$, $\pi/4$, $\pi/3$, $5\pi/12$, and $\pi/2$ are analysed herein. The proposed method was used to obtain the elastic buckling solution, and the buckling mode result q_c^h of the beam with crack damage was transformed into the dimensionless value $\overline{q}_c = q_c^h/q_{nc}$; the results are shown in Table 6.4. As shown, the values of $\overline{q}_c$ are all less than 1, which indicates that the buckling load values of the damaged beams are lower than those of the undamaged beams. Therefore, crack damage reduces the buckling load bearing capacity of curved beams. As the circularly subtended angle θ_c increases ($\theta_c = 0 \rightarrow \pi/2$) in the crack damage position, the value of $\overline{q}_c$ gradually decreases.

Buckling loads of curved beam with crack damage at different locations　　Table 6.4

θ_c	$\overline{q}_c$	θ_c	$\overline{q}_c$
0	0.999999	$\pi/3$	0.937817
$\pi/12$	0.994679	$5\pi/12$	0.924499
$\pi/6$	0.978168	$\pi/2$	0.919641
$\pi/4$	0.958574		

Similarly, the elastic buckling loads when the crack damage is located at the right half span of a curved beam with circular angle coordinates θ_c of $7\pi/12$, $2\pi/3$, $3\pi/4$, $5\pi/6$, and $11\pi/12$ can be obtained as well. The solutions of the elastic buckling load when the crack damage is located at each subtended angle of the full span were obtained, and the change curve is shown in Figure 6.6 (a). Owing to the symmetry of the model and the loading form, the beam has the same buckling load and minimum value at $\theta_c = \pi/2$ when the damage is located at the symmetric position of the left and right spans of the curved beam. Therefore, when a crack is located in the middle of the span, the elastic buckling limit load is minimal and instability failure can occur easily. Figure 6.6 (b) shows the differences in the curves of each buckling mode and non-buckling mode when the crack is located at the left half-span. As shown, the local position of the crack damage has a significant influence on the change in buckling mode, and the crack damage is the main factor that affects the change in the buckling mode. The adaptive final non-uniform mesh for solving $\theta_c = \pi/2$ is shown on the horizontal coordinate axis. As shown, a relatively dense mesh is used near the crack to adapt to the change in the buckling mode due to crack damage, which reflects the adaptability of the adaptive mesh method to changes in the buckling mode.

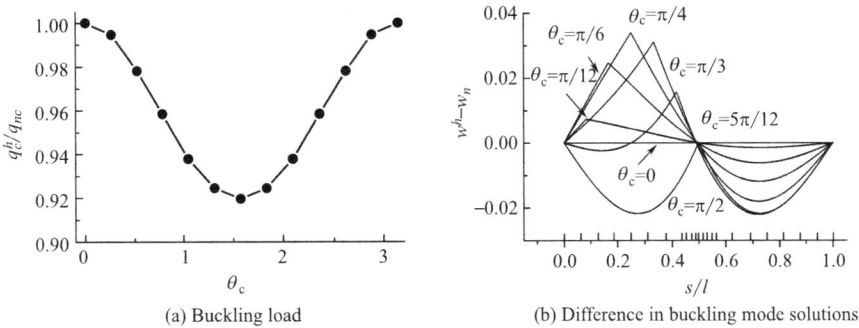

(a) Buckling load　　　　　　　　(b) Difference in buckling mode solutions

Figure 6.6　Elastic buckling of circularly curved beam with crack damage in different locations

6.5.4 Example 4: Different depths of single crack damage

The influence of the crack depth on the elastic buckling of the circularly curved beam was analysed. The fixed crack was located at the mid-span ($\theta_c = \pi/2$) and the elastic buckling was analysed when the crack damage size h_c/h was 0.1, 0.2, 0.3, 0.4, and 0.5. The solution of the elastic buckling load was obtained using the proposed method, and the results are listed in Table 6.5. As shown, as the crack damage $h_c(h_c/h = 0.1 \rightarrow 0.5)$ increased, $\bar{q}_c$ gradually decreased.

Buckling loads of curved beams with cracks of different magnitudes   Table 6.5

h_c/h	$\bar{q}_c$
0.1	0.995226
0.2	0.988219
0.3	0.976756
0.4	0.957391
0.5	0.919641

To observe and analyse the variation trend more intuitively, the change curve shown in Figure 6.7 (a) was plotted based on the buckling load results considering different crack depths. As shown, the buckling load has a quasi-linear decreasing trend with the increase in crack depth. When the crack extends to half the beam height $h_c/h = 0.5$, the buckling load bearing capacity decreased by approximately 10%. Figure 6.7 (b) shows the difference curves between each buckling mode and the undamaged buckling mode considering different crack damage sizes. As shown, the crack damage size has a significant influence on the change amplitude of the buckling mode. The higher the degree of crack damage, the higher the possibility of a change in the buckling mode. The horizontal axis in Figure 6.7 (b) depicts the adaptive non-uniform mesh obtained using the proposed method.

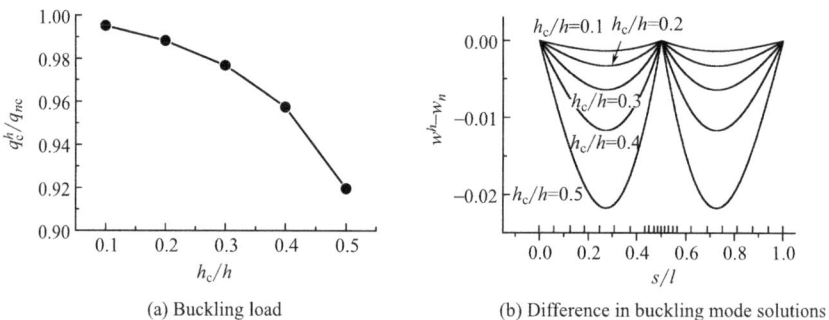

(a) Buckling load   (b) Difference in buckling mode solutions

Figure 6.7   Elastic buckling of circularly curved beam with different magnitudes of crack damage

6. 5. 5 Example 5: Different distributions of multiple crack damage

The influence of multi-crack damage on the elastic buckling of curved beams was analysed considering a circularly curved beam that is simply supported at both ends as shown in Figure 6. 8. The angle of the curved beam was $\theta = \pi$, and the other geometric and physical parameters were the same as those in Equation (6.12). The curved beam contained three cracks, and three working conditions were considered herein: Ⅰ (uniform distribution of crack damage along the curved beam), Ⅱ (crack damage concentrated at the end of the curved beam), and Ⅲ (crack damage concentrated in the middle of the curved beam). The coordinates of the crack distribution angles in each working condition are shown in Table 6.6, and the crack distribution is shown in Figure 6.8.

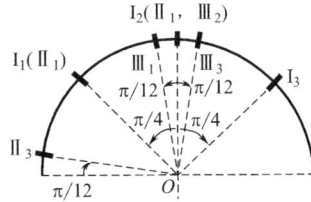

Figure 6. 8 Distributions of multiple cracks

The results of the multi-crack condition of the curved beamcomputed by the proposed method are shown in Table 6. 6. The buckling load value was least with working condition Ⅲ. Therefore, the denser the concentration of multi-crack damage in the middle of the span, the higher the possibility of crack-induced elastic instability. This is consistent with the conclusion obtained in Section 5. 3 that a single crack located in the middle of the span can cause higher elastic instability.

Buckling loads of curved beam with multiple cracks at different locations Table 6. 6

Cases	Location of the first crack	Location of the second crack	Location of the third crack	$\bar{q}_c$
Ⅰ	$\pi/4(\,Ⅰ_1)$	$\pi/2(\,Ⅰ_2)$	$3\pi/4(\,Ⅰ_3)$	0. 856826
Ⅱ	$\pi/4(\,Ⅱ_1)$	$\pi/2(\,Ⅱ_2)$	$\pi/12(\,Ⅱ_3)$	0. 881116
Ⅲ	$5\pi/12(\,Ⅲ_1)$	$\pi/2(\,Ⅲ_2)$	$7\pi/12(\,Ⅲ_3)$	0. 800923

The mesh distribution and buckling modes of circularly curved beams with multiple cracks are shown in Figure 6. 9. As shown, the local damage area of each crack has a significant influence on the change in the buckling mode, and multi-crack damage defects lead to a sharp change in the buckling mode. The proposed algorithm uses a relatively dense mesh in the regions near the crack damage to adapt to the changes in the buckling mode due to crack damage. Thus, an optimised non-uniform mesh is formed which ensures the reliability of the solution.

(a) Uniformly distributed cracks along curved beam

(b) Cracks concentrated at end of curved beam

(c) Cracks concentrated in middle of curved beam

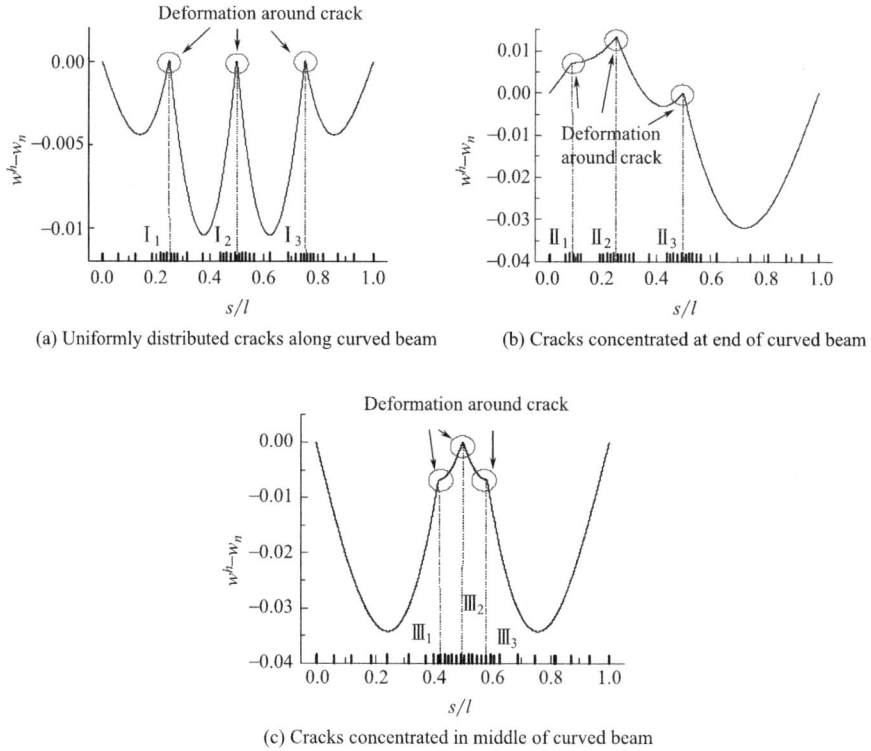

Figure 6. 9   Mesh distribution and buckling modes of circularly curved beam with multiple cracks

6. 6   Conclusions

The conclusions of this study are:

(1) An h-version adaptive finite element method based on the non-uniform Euler-Bernoulli beam theory was introduced herein to solve the elastic buckling problem of circularly curved beams with crack damage. A characterisation method for micro-crack damage in circularly curved beams was established, and the size (depth), location, and number of cracks were simulated and analysed. Adaptive mesh generation and the convergence of elastic buckling solutions were discussed. The effects of the subtended angle, damage location, number, and size of cracks on the elastic buckling load and buckling mode were analysed, and the effectiveness of the adaptive mesh generation method was verified.

(2) The finite element mesh adaptive analysis method can generate an optimised mesh, yielding high-precision buckling load and buckling mode solutions that satisfy the pre-specified error tolerance. The adaptive method uses a non-uniform distribution mesh, which prevents the element redundancy of a high-density uniform distribution mesh and improves the computational efficiency. The proposed adaptive algorithm divides the beam

into non-uniform meshes, with relatively dense meshes near the crack region to adapt to the changes in the buckling mode due to crack damage.

(3) Micro-crack damage is the main factor that induces vibrational disturbances, and the influence of crack damage was discussed quantitatively herein: the buckling load of beams with crack damage is lower than that of undamaged beams; crack damage reduces the buckling load bearing capacity of curved beams; crack damage induces a change in the buckling mode; the higher the degree of crack damage, the higher the possibility of a significant change in the buckling mode; and the higher the concentration of multiple cracks in the middle of the span, the higher the elastic instability.

The finite element mesh refinement method proposed herein improves the design process of the traditional finite element method: the mesh need not be refined at all positions and the crack region need not be accurately located, which significantly improves the efficiency and accuracy of the finite element method for computing elastic buckling problems. Herein, the research object was limited to the elastic buckling of circularly curved beams with crack damage, and no other cases were studied. In future studies, this method can be used to study the elastic buckling of complex models that are more consistent with reality. This can be achieved by adding more boundary conditions to the computation for the high-precision adaptive computation of the buckling loads of complex three-dimensional structures and solids. Furthermore, in the future, the error estimation and adaptive mesh refinement methods may be extended to fine numerical models and the high-precision computation of general structural eigenvalue problems (displacement field) and solid stress (displacement derivative field).

References

[1] Matallah M, La B C, Maurel O A, et al. Practical method to estimate crack openings in concrete structures [J]. International Journal for Numerical and Analytical Methods in Geomechanics, 2010, 34 (15): 1615-1633.

[2] Khiem N T, Toan L K. A novel method for crack detection in beam-like structures by measurements of natural frequencies [J]. Journal of Sound and Vibration, 2014, 333 (18): 4084-4103.

[3] Jassim Z A, Ali N N, Mustapha F, et al. A review on the vibration analysis for a damage occurrence of a cantilever beam [J]. Engineering Failure Analysis, 2013, 31: 442-461.

[4] Yang J, Chen Y. Free vibration and buckling analyses of functionally graded beams with edge cracks [J]. Composite Structures, 2008, 83 (1): 48-60.

[5] Wang Y L, Ju Y, Chen J L, et al. Adaptive finite element-discrete element analysis for the multistage supercritical CO_2 fracturing of horizontal wells in tight reservoirs considering pre-existing fractures and thermal-hydro-mechanical coupling [J]. Journal of Natural Gas Science and Engineering, 2019, 61: 251-269.

[6] Piovan M T, Domini S, Ramirez J M, et al. In-plane and out-of-plane dynamics and buckling of functionally graded circularly curved beams [J]. Composite Structures, 2012, 94 (11): 3194-3206.

[7] Yuan T, Hai Q. Elastic Buckling and Free Vibration Analysis of Functionally Graded Timoshenko Beam with Nonlocal Strain Gradient Integral Model [J]. Applied Mathematical Modelling, 2021, 96: 657-677.

[8] Cerri M N, Dilena M, Ruta G C, et al. Vibration and damage detection in undamaged and cracked circularly arches: experimental and analytical results [J]. Journal of Sound and Vibration, 2008, 314 (1-2): 83-94.

[9] Timoshenko S P, Gere J M. Theory of Elastic Stability, 2nd Edition, [M]. NewYork: McGraw-Hill, 1961.

[10] Caddemi S, Caliò I, Cannizzaro F, et al. The influence of multiple cracks on tensile and compressive buckling of shear deformable beams [J]. International Journal of Solids and Structures, 2013, 50 (20-21): 3166-3183.

[11] Khatri A P Katikala, S R, Kotapati V K, et al. Effect of load height on elastic buckling behavior of I-shaped cellular beams [J]. Structures, 2021, 33: 1923-1935.

[12] Kim M Y, Kim, N I Min B C, et al. Analytical and numerical study on spatial free vibration of non-symmetric thin-walled curved beams [J]. Journal of Sound and Vibration, 2002, 258 (4): 595-618.

[13] Sandeep A S. Out of plane free vibration analysis of circular curved beam with variable cross section by spectral element method [D]. National Institute of Technology Rourkela, 2017.

[14] Poojary J, Roy S K. In Plane Radial Vibration of Uncracked and Cracked Circular Curved Beams Subjected to Moving Loads [J]. International Journal of Structural Stability and Dynamics, 2021, 21 (10): 2510146.

[15] Karaagac C, Öztürk H, Sabuncu M, et al. Crack effects on the in-plane static and dynamic stabilities of a curved beam with an edge crack [J]. Journal of Sound and Vibration, 2011, 330 (8): 1718-1736.

[16] Kim K H, Kim J H. Effect of a crack on the dynamic stability of a free-free beam subjected to a follower force [J]. Journal of Sound and Vibration, 2000, 233 (1): 119-135.

[17] Radhakrishnan V M. Vibration of a simply supported beam with an edge crack [J]. Journal of the Institution of Engineers (India), Part MM: Metallurgy and Material Science Division, 2001, 82 (1): 29-32.

[18] Taylan D M, Ay E Y. Experimental modal analysis of curved composite beam with transverse open crack [J]. Journal of Sound and Vibration, 2018, 436 (8): 155-164.

[19] Isa A, Rahman Z A, Assadi H A, et al. The effect of crack depth on the dynamic characteristics of homogeneous beam [C]. International Conference on Science and Social Research, 2010, IEEE.

[20] Afsar A M, Sekine H. Crack spacing effect on the brittle fracture characteristics of semi-infinite functionally graded materials with periodic edge cracks [J]. International Journal of Fracture, 2000, 102 (3): 61-66.

[21] Jamia N, El-Borgi S, Fernandes R, et al. Analysis of an arbitrarily oriented crack in a functionally graded plane using a non-local approach [J]. Theoretical and Applied Fracture Mechanics, 2016, 85: 387-397.

[22] Nikhil Y, Jeyashree T M. DYnamic response of a cracked beam under free vibration [J]. International Journal of Civil Structural Environmental and Infrastructure Engineering Research and Development, 2016, 6 (2): 43-56.

[23] Sutar K M, Pattnaik S. Vibration characteristics of a cracked cantilever beam under free vibration [J]. Noise and Vibration Worldwide, 2010, 41 (9): 16-21.

[24] Sun Z, Wang D, Li X, et al. Isogeometric Free Vibration Analysis of Curved Euler-Bernoulli Beams

with Particular Emphasis on Accuracy Study [J]. International Journal of Structural Stability and Dynamics, 2020, 21 (01): 2510011.

[25] Chang K C, Kim C W. Modal-parameter identification and vibration-based damage detection of a damaged steel truss bridge [J]. Engineering Structures, 2016, 122: 156-173.

[26] Fan W, Qiao P. Vibration-based damage identification methods: a review and comparative study [J]. Structural Health Monitoring, 2011, 10 (1): 83-111.

[27] Yang Y, Yang J. State-of-the-art review on modal identification and damage detection of bridges by moving test vehicles [J]. International Journal of Structural Stability and Dynamics, 2018, 18 (02): 1850025.

[28] Wang Y, Ju Y, Zhuang Z, et al. Adaptive finite element analysis for damage detection of non-uniform Euler-Bernoulli beams with multiple cracks based on natural frequencies [J]. Engineering Computations, 2018, 35 (3): 1203-1229.

[29] Flugge. Stresses in shells [M]. Springer, 1973.

[30] Leissa A W, Nordgren R P. Vibration of Shells [J]. Journal of Applied Mechanics, 1993, 41 (2): 544.

[31] Wu J S, Chiang L K. Free vibration of a circularly curved Timoshenko beam normal to its initial plane using finite curved beam elements [J]. Computers and structures, 2004, 82 (29/30): 2525-2540.

[32] Ojah P K. Free vibration analysis of circular curved beam by spectral element method [D]. National Institute of Technology Rourkela. , 2015.

[33] Yousefi A, Rastgoo A. Free vibration of functionally graded spatial curved beams [J]. Composite Structures, 2011, 93 (11): 3048-3056.

[34] Chondros T G, Dimarogonas A D, Yao J, et al. Vibration of a beam with a breathing crack [J]. Journal of Sound and Vibration, 2001, 239 (1): 57-67.

[35] Rezaee M, Fekrmi. A theoretical and experimental investigation on free vibration behavior of a cantilever beam with a breathing crack [J]. Shock and vibration, 2012, 19 (2): 175-1872 (1698).

[36] Arndt M, Machado R D, Scremin A, et al. An adaptive generalized finite element method applied to free vibration analysis of straight bars and trusses [J]. Journal of Sound and Vibration, 2010, 329 (6): 659-672.

[37] Karaagac C, Öztürk H, Sabuncu M, et al. Free vibration and lateral buckling of a cantilever slender beam with an edge crack: experimental and numerical studies [J]. Journal of Sound and Vibration, 2009, 326 (1-2): 235-250.

[38] Kishen J M C, and Kumar A. Finite element analysis for fracture behavior of cracked beam-columns [J]. Finite Elements in Analysis and Design, 2004, 40 (13-14): 1773-1789.

[39] Melenk J M, Wohlmuth B I. On residual-based a posteriori error estimation in hp-FEM [J]. Advances in Computational Mathematics, 2001, 15 (1-4): 311-331.

[40] Yuan S, Wang Y, Ye K, et al. An adaptive FEM for buckling analysis of non-uniform Bernoulli-Euler members via the element energy projection technique [J]. Mathematical Problems in Engineering, 2013, 40 (7): 221-239.

[41] Wang Z Q, Huang L H, Li X B, et al. Vibration analysis of circular cylindrical shells under arbitrary boundary conditions [J]. Ship Science and Technology, 2017, 39 (4): 24-29.

[42] Wang Y. An h-version adaptive FEM for eigenproblems in system of second order ODEs: Vector Sturm-Liouville problems and free vibration of curved beam [J]. Engineering Computations, 2020, 37 (1): 1210-1225.

[43] Wang Y, Wang J. An *hp*-version adaptive finite element algorithm for eigensolutions of free vibration of moderately thick circular cylindrical shells via error homogenization and higher-order interpolation [J]. Engineering Computations, 2021.

[44] Wang Y. Adaptive analysis of damage and fracture in rock with multiphysical fields coupling [J]. Springer Press, 2021.

[45] Petrov E, Geradin M. Finite element theory for curved and twisted beams based on exact solutions for three-dimensional solids Part 2: Anisotropic and advanced beam models [J]. Computer Methods in Applied Mechanics and Engineering, 1998, 165 (1/4): 93-127.

[46] Lebsack M J. Elasticity-based vibrations of hollow anisotropic beams and an evaluation of the shape factor for hollow anisotropic sections [D]. Colorado: Colorado State University. , 2012.

[47] Calik-Karakose U H, Orakdogen E, Saygun A I, et al. A curved shell finite element for the geometrically non-linear analysis of box-girder beams curved in plan [J]. Structural Engineering and Mechanics, 2014, 52 (2): 221-238.

[48] Park S W, Fujii D, Fujitani Y, et al. Finite element analysis of thin-walled beam considering shear warping deformation: Bending-torsional analysis by using nonuniform shear warping function [J]. Journal of Structural and Construction Engineering, 1996, 61 (484): 65-74.

[49] Qi H, Wang Z, Zhang Z, et al. An efficient finite element for restrained torsion of thin-walled beams including the effect of warping and shear deformation [J]. IOP Conference Series Earth and Environmental Science, 2019, 233 (3): 32029.

[50] Zienkiewicz O C. The background of error estimation and adaptivity in finite element computations [J]. Computer Methods in Applied Mechanics and Engineering, 2006, 195 (4-6): 207-213.

[51] Bespalov A, Haberl A, Praetorius D, et al. Adaptive FEM with coarse initial mesh guarantees optimal convergence rates for compactly perturbed elliptic problems [J]. Computer Methods in Applied Mechanics and Engineering, 2017, 317: 318-340.

[52] Arthurs C J, Bishop M J, Kay D, et al. Efficient simulation of cardiac electrical propagation using high-order finite elements II: adaptive *p*-version [J]. Journal of Computational Physics, 2013, 253: 443-470.

[53] Gomez-Revuelto I, Garcia-Castillo L E, Llorente-Romano S, et al. A three-dimensional self-adaptive hp finite element method for the characterization of waveguide discontinuities [J]. Computer Methods in Applied Mechanics and Engineering, 2012, 249: 62-74.

[54] Krishnan A, Suresh Y J. A simple cubic linear element for static and free vibration analyses of curved beams [J]. Computers and Structures, 2003, 68: 473-489.

[55] Abu-Hilal M. Forced vibration of Euler-Bernoulli beams by means of dynamic Green functions [J]. Journal of Sound and Vibration, 2010, 267 (2): 191-207.

[56] Zienkiewicz O C, Taylor R L, Zhu J Z, et al. The finite element method: its basis and fundamentals (7th Edition) [J]. Oxford: Elsevier (Singapore)Pte Ltd. , 2015.

[57] Wiberg N E, Bausys R, Hager P, et al. Adaptive *h*-version eigenfrequency analysis [J]. Computers and Structures, 1999, 71 (5): 565-584.

[58] Wiberg N E, Bausys R, Hager P, et al. Improved eigenfrequencies and eigenmodes in free vibration analysis [J]. Computers and Structures, 1999, 73 (1-5): 79-89.

[59] Clough R W, Penzien J. Dynamics of structures, 2nd Edition, [M]. New York: McGraw-Hill, 1993.

[60] Tripathy B, Rao K P. Curved composite beams — optimum lay-up for buckling by ranking [J]. Computers and Structures, 1991, 41 (1): 75-82.

Chapter 7
Adaptive finite element method for vibration of cylindrical shells

7. 1   Introduction

The dynamic analysis of structures and elastomers as isolation systems is an important basis for the study of seismic-resistant structures and rock mass-induced earthquakes[1-5], and cylindrical shells, which are supporting structures or storage cavities, are widely used in structural engineering, rock mass engineering, and aerospace engineering. In studying the vibration, instability, and buckling of structures, the dynamic characteristics are essential for understanding the failure behaviour[6-11]. Currently, the conventional thin shell theory[12-14], which is based on the Kirchhoff-Love hypothesis ignoring transverse shear deformation, is often used to study shell problems[15-19], and for shell structures with small shear stiffness (i. e. prone to significant transverse shear deformation), certain errors will be introduced[20-23]. In addition, the increase in wall thickness of the plate and shell structure often results in exceedance of the application scope of the thin-wall theory in practical engineering, and the influence of transverse shear deformation must be considered. Compared with the free vibration theory of thin shells, that of moderately thick shells considers the influence of transverse shear deformation and rotational inertia, which makes the solution more reliable. At the same time, the moderately thick cylindrical shell theory can take into consideration different wall thicknesses, lengths, circumferential wave numbers and boundary conditions, which have significant impacts on the computation of frequencies of moderately thick circular cylindrical shells.

Hence, the free vibration of moderately thick cylindrical shells, which is a typical eigenvalue problem of second-order ordinary differential equations, is studied. It is difficult to solve the eigenvalue problem of ordinary differential equations for the continuous order frequencies and vibration modes of moderately thick cylindrical shells with various boundary conditions, different circumferential wave numbers and thickness-length ratios. At present, there are many studies on variable wall thicknesses, lengths[24-26], circumferential wave numbers, and boundary conditions[27-29]. These studies have conducted systematic and specific analyses of the aforementioned influencing factors. It can be said that the current research on these factors has been adequate. However, most studies are generally based on the thick-wall or thin-wall theory, and only few studies have evaluated these factors based on the moderately thick shell theory. It is necessary to take into consideration the transverse shear deformation and moment of inertia when analysing these factors using the moderately thick shell free vibration theory; thus, employing this theory is difficult. We aim to tackle this relatively difficult subject and attempt to make this theory more accurate, realistic, and reliable compared with the conventional shell theories. Therefore, we analyse the eigenvalue problems of complex geometric conditions using the moderately thick shell theory.

The finite element method is commonly used to solve the free vibration analysis of rotating cylindrical shells with complex structural forms and boundary conditions, and the numerical solution of high-precision natural vibration frequency and vibration mode has become the main basis for accurate damage identification of structures[30-31]. Compared with the conventional finite element method, the adaptive finite element method can obtain a high-precision solution satisfying the user-specified error tolerance[32], and the adaptive algorithm has become an important method for optimising the mesh and improving the solution accuracy[33, 34]. Considering the intermediate results of some variables of the finite element model as error estimates during adaptive analysis, error analysis is carried out to explore the real error distribution. This can provide guidance on the density settings of the mesh and nodes and thereby effectively improve the precision of the finite element simulation[35]. The analysis mainly includes the p-version adaptive method for improving cell order[36], h-version adaptive method for increasing mesh density[31, 37, 38], and hp-version adaptive method combining the two[39, 40]. The adaptive finite element method exhibits good potential in solving structural eigenvalue problems such as free vibration and elastic stability of beams, plates, and shells with high accuracy[41-43], and it can also tackle iteration problems[44].

To solve the above-mentioned vibration problems using the adaptive finite element method, the algorithm of free vibration analysis of Euler-Bernoulli beams with variable cross-sections established earlier[45, 46] is extended. The displacement superconvergent patch recovery method, high-order shape function interpolation technology, energy mode error estimation technology, and mesh subdivision and refinement method are introduced[47]. A set of h-version finite element adaptive solutions for the free vibration analysis of rotating cylindrical shells[48, 49] is presented. The good computational efficiency and stability of the proposed method are verified by numerical examples. Thus, the user-specified error tolerances of multiple models can be perfectly met. The algorithm in this study can be extensively applied to various mechanics problems such as vibration, wave, and wall thickness[50]. Moreover, this type of adaptive finite element method can be further developed for the problem of rotating cylindrical shells.

In thischapter, we provide the equations for the free vibration of a moderately thick cylindrical shell by listing some differential governing equations in Section 7. 2. Then, the objectives and steps of the adaptive solution for the free vibration of a moderately thick cylindrical shell are presented in Section 7. 3. The expressions of the finite element solution are provided in Section 7. 4. Subsequently, the error estimation of the algorithm is discussed in Sections 7. 5 and 7. 6, and the schemes for the mesh subdivision and refinement are briefly illustrated. In Section 7. 7, the accuracy and applicability of the algorithm are verified by analysing some numerical examples. Finally, the conclusions are summarised in Section 7. 8.

7.2 Partial differential governing equations for the free vibration of rotating cylindrical shells

The moderately thick cylindrical shell, as shown in Figure 7.1, is considered, where ox is the rotation axis. The local coordinate system $\alpha\beta\gamma$ of point A on the middle surface is established, where α is along the tangential direction of the meridian, β along the tangential direction of the latitudinal circle, and γ along the normal direction. The five independent displacements of vibration for the moderately thick cylindrical shell are the linear displacements u along α, v along β, and w along γ, as well as angular displacements ϕ and ψ around α and around β, respectively. The symbols r, h, κ, J, and l denote the centre surface radius of the cylindrical shell, section thickness, shear stiffness correction coefficient, moment of inertia, and length, respectively. The elastic modulus of the material, shear modulus, Poisson's ratio, and density are symbolised as E, G, μ, and ρ, respectively.

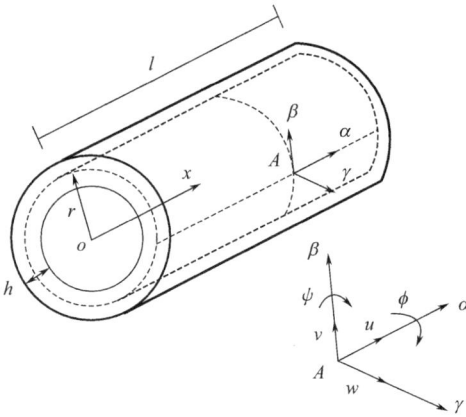

Figure 7.1 Coordinate system and symbols for the moderately thick circular cylindrical shell

In this study, the differential governing equation for the free vibration of the moderately thick cylindrical shell is as follows:

$$(1/2)K\left[(1-\mu)n^2u/r^2-2u''-(1+\mu)nv'/r-2\mu w'/r\right]=\omega^2\rho hu \qquad (7.1a)$$

$$(1/2)K\left[(1+\mu)nu'/r+2n^2v/r^2-(1-\mu)(1+h^2/12r^2)v''+2nw/r^2\right.$$
$$\left.+(1-\mu)nh^2\phi'/(12r^2)-(1-\mu)h^2\psi''/(12r)\right]+\bar{\kappa}Gh(v/r^2+nw/r^2-\psi/r)$$
$$=\omega^2\rho hv \qquad (7.1b)$$

$$K(\mu u'/r+nv/r^2+w/r)+\bar{\kappa}Gh(n^2w/r^2+nv/r^2-n\phi/r-w''-\phi')$$
$$=\omega^2\rho hw \qquad (7.1c)$$

$$(h^2/24)K\left[-(1-\mu)nv'/r^2+(1-\mu)n^2\phi/r^2-2\phi''-(1+\mu)n\psi'/r\right]+\bar{\kappa}Gh(w'+\phi)$$
$$=\omega^2\rho J\phi \qquad (7.1d)$$

$$(h^2/24)K\left[-(1-\mu)v''/r+(1+\mu)n\phi'/r+2n^2\psi/r^2-(1-\mu)\psi''\right]$$
$$+\bar{\kappa}Gh(\psi-v/r-nw/r)=\omega^2\rho J\psi \qquad (7.1e)$$

where $()''=\mathrm{d}^2()/\mathrm{d}x^2$, $()'=\mathrm{d}()/\mathrm{d}x$, ω is the natural frequency, and n is the wave number of the circumferential vibration mode[51].

The eigenvalue problem of the second-order ordinary differential equations can be expressed in the following matrix form:

$$\boldsymbol{Lu}=\omega^2\boldsymbol{Ru} \qquad (7.2)$$

where $\boldsymbol{L}$ is the corresponding differential operator matrix, and $\boldsymbol{u}$ is the mode (displacement) function vector.

$$\boldsymbol{u} = \{u,\ v,\ w,\ \phi,\ \psi\}^{\mathrm{T}} = \{u_1,\ u_2,\ u_3,\ u_4,\ u_5\}^{\mathrm{T}} \tag{7.3}$$

where ω and $\boldsymbol{u}$ correspond to the eigenvalue and eigenvector, respectively, and $(\omega,\ \boldsymbol{u})$ is also collectively called an eigenpair in this study.

7.3　Mesh refinement procedure of finite element method

We set an eigenpair of a certain order for the free vibration of a rotating cylindrical shell required to be solved in Equation (7.2) as $(\omega,\ \boldsymbol{u})$. In this study, the goal of the free vibration of a thick cylindrical shell for adaptation is as follows: in the case of an unknown exact solution $(\omega,\ \boldsymbol{u})$ and given error tolerance Tol in advance, an optimised finite element mesh π is sought, where the finite element solutions $(\omega^h,\ \boldsymbol{u}^h)$ on the mesh simultaneously satisfies

$$|\omega - \omega^h| \leqslant Tol \cdot (1 + |\omega|) \tag{7.4}$$

$$\max_{0 < x < l} |u_i(x) - u_i^h(x)| \leqslant Tol, \quad \max_i (\max_{0 < x < l} |u_i(x)|) = 1,\ i = 1,\ 2,\ \cdots,\ 5 \tag{7.5}$$

When implemented, because the exact solution $(\omega,\ \boldsymbol{u})$ is not known in advance, the above objective cannot be used as the stopping criterion. As for the mode of vibration, $\boldsymbol{u}^*$, which is the superconvergent patch recovery solution of the mode of vibration (displacement) described below, has a higher convergence order than $\boldsymbol{u}^h$, that is, $\boldsymbol{u}^*$ is closer to the exact solution $\boldsymbol{u}$ than $\boldsymbol{u}^h$. Therefore, $\boldsymbol{u}^*$ is used to estimate and control the error of $\boldsymbol{u}^h$ instead of $\boldsymbol{u}$ to formulate the error estimation in the form of an energy mode. For the frequency, the displacement superconvergence solution is used to compute the Rayleigh quotient[52] and obtain the superconvergence solution. The frequency superconvergence solution obtained by the Rayleigh equation has a higher convergence order than the vibration-mode superconvergence solution. In this study, the error of the vibration mode solution is estimated and controlled to ensure that the frequency solution satisfies the error requirements, thus forming the stopping criterion of the method by controlling the vibration mode error. By subdividing and refining the mesh based on the error estimation of the finite element solution, the optimised mesh and high-precision frequency and mode solutions satisfying the error tolerances can be obtained, forming a set of h-version finite element adaptive schemes. The adaptive solution steps include the following:

（1）**Finite element solution of rotating cylindrical shells.** The conventional finite element computation is performed on the current mesh (the initial mesh is provided by the user) and the finite element solution under the current mesh is obtained using the inverse and subspace iteration techniques.

（2）**Error estimation based on superconvergent vibration modes.** The superconvergence solution of frequency and mode under the current mesh can be obtained using the superconvergence algorithm of finite element post-processing. Meanwhile, the error estimates between the superconvergent solution and finite element solution are obtained. Hence, the errors can be controlled using the superconvergent patch recovery solutions of vibration modes

(3) **Element subdivision and mesh refinement.** For the element whose error estimation does not meet the error tolerance, the mesh subdivision and refinement method is used to subdivide it. Then, a new finite element mesh is obtained and the first step is repeated. If all elements meet the error tolerances, mesh subdivision is not required, and the computation ends. In this study, the following problems are considered to test the effectiveness of the mesh refinement procedure of the finite element method: thin-wall thickness, moderately thick-wall thickness, different circumferential wave numbers and thickness-to-length ratios, and hinged and clamped boundary conditions.

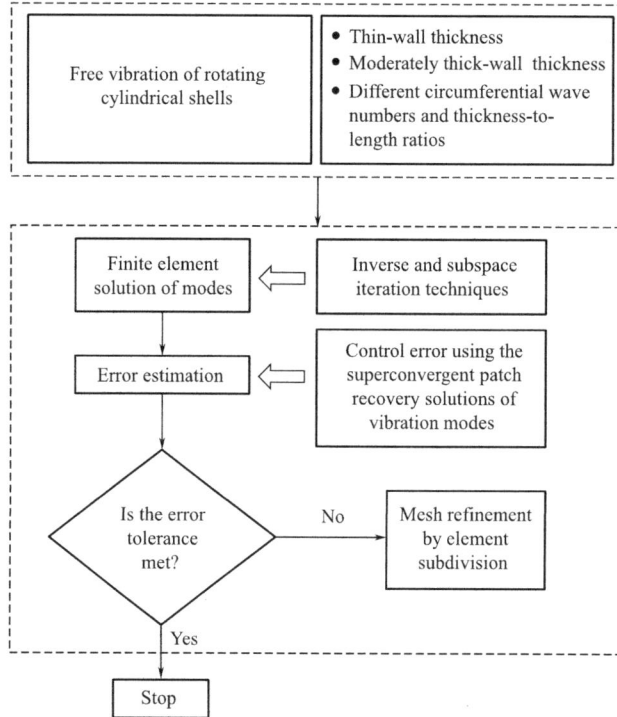

Figure 7. 2   Procedure of mesh refinement of finite element method for rotating cylindrical shells based on superconvergent vibration modes

7. 4   Finite element discretisation

To solve the free vibration problem of a moderately thick cylindrical shell in Equation (7. 2), given the finite element mesh π, the following linear matrix eigenvalue problem is established using the conventional finite element method:

$$\boldsymbol{Ku} = \omega^2 \boldsymbol{Mu} \tag{7.6}$$

where $\boldsymbol{K}$ and $\boldsymbol{M}$ are the static stiffness matrix and uniform mass matrix, respectively. The finite element solution (ω^h, $\boldsymbol{u}^h$) can be obtained by solving the current mesh. It should be noted that the accuracy of the solution is dependent on the mesh, and a high-precision solution requires a high-quality finite element mesh.

7.5　Error estimation

The displacement-type finite element method has superconvergence in displacement onthe node and has a higher convergence order and accuracy than the method using the internal points of the element[53, 54]. Using the high-precision value on the high-order node and the high-order shape function for interpolation, the superconvergence solution for the displacement of each element can be obtained[31, 38, 40]. In this study, for the free vibration of a moderately thick cylindrical shell, the finite element solution of the modes (displacement) can also be obtained. By combining the adjacent elements, the interpolation form for the displacement superconvergence solution of the element is

$$u_i^* (x) = \boldsymbol{P}\boldsymbol{a}, \quad i = 1, 2, \cdots, 5 \tag{7.7}$$

where $\boldsymbol{P}$ is the given function vector and $\boldsymbol{a}$ is the vector of undetermined coefficients.

$$\boldsymbol{P} = \begin{bmatrix} 1 & x & \cdots & x^p \end{bmatrix}, \quad \boldsymbol{a} = \begin{bmatrix} a_1 & a_2 & \cdots & a_m \end{bmatrix}^{\mathrm{T}} \tag{7.8}$$

The value of coefficient $\boldsymbol{a}$ is determined by obtaining the minimum value of the following function so that the result in Equation (7.7) is equal to the current node displacement value at the element node:

$$\Pi = \sum_{j=1}^{n} (u_i^* (x_j) - \boldsymbol{P}(x_j)\boldsymbol{a}), \quad i = 1, 2, \cdots, 5 \tag{7.9}$$

where n is the number of nodes for assembling elements.

Using the leastsquares method to solve Equation (7.9), the value of coefficient $\boldsymbol{a}$ can be obtained as

$$\boldsymbol{a} = \boldsymbol{A}^{-1}\boldsymbol{b} \tag{7.10}$$

where each coefficient matrix is

$$\begin{aligned} \boldsymbol{A} &= \sum_{j=1}^{n} \boldsymbol{P}(x_j)^{\mathrm{T}}\boldsymbol{P}(x_j) \\ \boldsymbol{b} &= \sum_{j=1}^{n} \boldsymbol{P}(x_j)^{\mathrm{T}}u_i^h(x_j), \quad i = 1, 2, \cdots, 5 \end{aligned} \tag{7.11}$$

After determining $\boldsymbol{a}$, the displacement superconvergence solution of the splice element can be obtained using Equation (7.7). The superconvergence solution has a higher convergence order than the finite element solution under the current mesh, which can be used to estimate the error of the finite element solution.

Using the obtained displacementsuperconvergence solution and by calculating the Rayleigh quotient as follows, the frequency superconvergence solution ω^* can be obtained:

$$\omega^* = \sqrt{\frac{a(\boldsymbol{u}^*, \boldsymbol{u}^*)}{b(\boldsymbol{u}^*, \boldsymbol{u}^*)}} \qquad (7.12)$$

where $a(\ )$ and $b(\ )$ are the inner products of the strain energy and kinetic energy, respectively. Using the displacement superconvergence solution and finite element solution under the current mesh, the error estimation in the form of an energy mode can be obtained[55]:

$$\|\boldsymbol{e}^*\| \leqslant Tol \cdot [(\|\boldsymbol{u}^h\|^2 + \|\boldsymbol{e}^*\|^2)/n_e]^{1/2} \qquad (7.13)$$

where n_e is the number of assembling elements, and $\boldsymbol{e}^*$ is the error of displacement superconvergence solution $\boldsymbol{u}^*$ and finite element solution $\boldsymbol{u}^h$, which has an energy norm expression as follows:

$$\|\boldsymbol{e}^*\| = [a(\boldsymbol{e}^*, \boldsymbol{e}^*)]^{1/2} = \left[\int_0^l \boldsymbol{e}^{*T} \boldsymbol{L} \boldsymbol{e}^* \, \mathrm{d}x\right]^{1/2}, \quad \boldsymbol{e}^* = \boldsymbol{u}^* - \boldsymbol{u}^h \qquad (7.14)$$

Furthermore, Equation (7.13) can be written as

$$\xi = \frac{\|\boldsymbol{e}^*\|}{\bar{e}}, \quad \bar{e} = Tol \cdot [(\|\boldsymbol{u}^h\|^2 + \|\boldsymbol{e}^*\|^2)/n_e]^{1/2} \qquad (7.15)$$

where ξ is the relative error, which should be satisfied:

$$\xi \leqslant 1 \qquad (7.16)$$

It should be pointed out that the frequency superconvergence solution computed by the Rayleigh quotient has a higher convergence order than the vibration mode superconvergence solution. By error estimation and control of the vibration mode, a high-precision vibration mode can be obtained, and frequency accuracy can be ensured.

7.6　Element subdivision and mesh refinement

If the error estimation (Equation (7.16)) on the current element e is not satisfied, it means that the element needs subdivision and refinement to reduce the solution error. In this study, the mesh subdivision and refinement method is adopted to evenly subdivide the current element into multiple elements. The length of the new element is related to the error in the current element and the order of the element. The estimation formula is

$$h_{new} = \xi^{-1/m} h_{old} \qquad (7.17)$$

where h_{new} is the length of the new element, and h_{old} is the length of the current cell e. To make the adaptive computation more efficient and avoid mesh redundancy, in this study, the number of each cell subdivision and refinement is controlled, and the implementation scheme is as follows:

$$n_{new} = \min(\lfloor \xi^{1/m} \rfloor, d) \qquad (7.18)$$

where n_{new} is the number of new cells after mesh subdivision, d is the maximum number of element subdivisions, and $\lfloor \cdot \rfloor$ denotes the rounded down symbol.

After the above finite element solution error estimation and mesh subdivision and refinement, theoptimised mesh can be finally achieved, and the high-precision frequency and mode solutions satisfying the error tolerances on the optimised mesh can be obtained.

7. 7 Numerical examples

The algorithm presented inthis study was implemented into a Fortran 90 program and applied to numerical examples for solving the free vibration of a variety of moderately thick cylindrical shells using the proposed method. The accuracy of the computation results of the proposed method was verified by comparing the results of typical moderately thick shells and thin shells. Furthermore, the applicability of the method for solving the free vibration problems of moderately thick cylindrical shells under different circumferential wave numbers, thickness-to-length ratios, and boundary conditions was evaluated. All the computation examples presented in this section adopted cubic element, and the initial mesh adopted two elements. The given initial error tolerance was $Tol = 10^{-4}$, and the maximum number of element subdivisions was set as $d = 6$. In this study, k represents the order and is indicated as a subscript of the frequency or mode symbol.

7. 7. 1 Example 1: Thin-wall thickness

First, the free vibration of a moderately thick cylindrical shell simply supported at both ends was considered using the following basic parameters:

$$r = 148.234 \text{ mm}, \ h = 0.508 \text{ mm}, \ l = 298.2 \text{ mm}, \ J = h^3/12,$$
$$E = 203.5 \text{ GPa}, \ \kappa = 5/6, \ \mu = 0.285, \ \rho = 7846 \text{ kg/m}^3 \tag{7.19}$$

The proposed method was used to solve the firsttwo orders ($k = 1, 2$) of eigenpairs with different circumferential wave numbers ($n = 1\text{-}15$). Table 7. 1 presents the computed frequency results. Sivadas et al.[56] and Bespalov et al.[34] used the moderately thick shell theory to solve this problem by using the mixed finite element method and dynamic stiffness method, respectively. It can be seen from Table 7. 1 that the results of this study are in good agreement with those of Sivadas et al.[56] and Chen and Ye[57], which verifies the reliability of the proposed method to solve the frequency of each order.

Frequencies ω (Hz) of the moderately thick circular cylindrical shell　　　Table 7. 1

n	Frequency values					
	$k=1$			$k=2$		
	ω_1^h	$\omega_1^{h\,①}$	$\omega_1^{h\,②}$	ω_2^h	$\omega_2^{h\,①}$	$\omega_2^{h\,②}$
1	3270. 83	3270. 6	3270. 9	4838. 37	4837. 8	4838. 3
2	1862. 15	1862. 0	1862. 1	3725. 47	3725. 1	3725. 4
3	1101. 82	1101. 8	1101. 8	2742. 94	2742. 8	2742. 9
4	705. 735	705. 9	705. 7	2018. 22	2018. 2	2018. 2
5	497. 478	497. 5	497. 4	1515. 04	1515. 2	1515. 0

continued

n	Frequency values					
	$k=1$			$k=2$		
	ω_1^h	$\omega_1^{h\,①}$	$\omega_1^{h\,②}$	ω_2^h	$\omega_2^{h\,①}$	$\omega_2^{h\,②}$
6	400. 081	400. 1	400. 0	1174. 91	1175. 1	1174. 8
7	380. 712	380. 7	380. 6	953. 569	953. 7	953. 5
8	416. 733	416. 7	416. 6	824. 181	824. 3	824. 1
9	488. 578	488. 6	488. 5	770. 262	770. 4	770. 2
10	583. 826	583. 8	583. 8	778. 263	778. 3	778. 0
11	696. 132	696. 1	696. 1	833. 986	834. 1	833. 9
12	822. 552	822. 5	822. 5	925. 245	925. 3	925. 1
13	961. 631	961. 6	961. 6	1042. 66	1042. 7	1042. 5
14	1112. 81	1112. 7	1112. 8	1180. 34	1180. 3	1180. 2
15	1275. 74	1275. 8	1275. 6	1334. 63	1334. 5	1334. 5

Source: ①Results from paper [56]; ②Results from paper [57].

Figures 7. 3-7. 6 illustrate the first mode of $n=1$ and $n=10$, with the final mesh distribution of the adaptive solution marked on the horizontal axis. It should be noted that, for convenience of intuitive display and comparative analysis, the vibration mode results are normalised (the maximum vibration mode value is 1), and the horizontal axis is normalised in the vibration mode diagram (the horizontal axis is x/l). As depicted in Figures 7. 3 and 7. 4, for the first mode at $n=1$, the changes are somewhat drastic at both ends, and the adaptive method divides a relatively dense mesh. The mode of vibration is relatively even in the middle region, and only a sparse mesh is required.

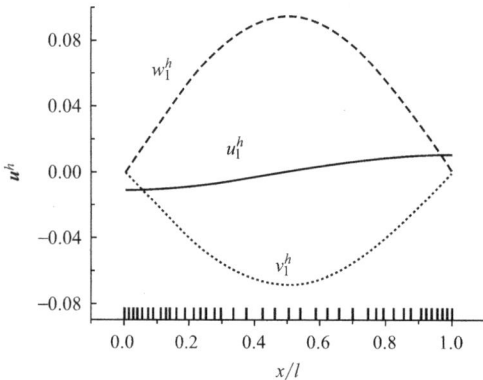

Figure 7. 3 Vibration mode (u_1^h, v_1^h, and w_1^h) and adaptive mesh with $n=1$

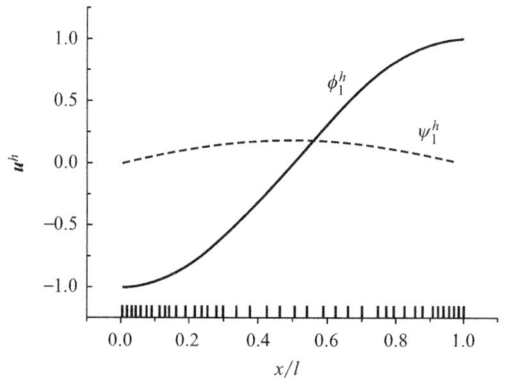

Figure 7. 4 Vibration mode (ϕ_1^h and ψ_1^h) and adaptive mesh with $n=1$

As can be seen fromFigures 7. 5 and 7. 6, for the first mode at $n=10$, the change is relatively gentle in both domains, and the adaptive method can divide a relatively sparse and uniform mesh. The finite element mesh adaptively refines and optimises according to the change in vibration modes, indicating its applicability.

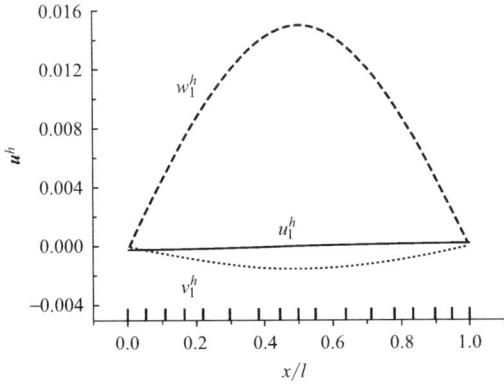

Figure 7.5   Vibration mode (u_1^h, v_1^h, and w_1^h) and adaptive mesh with $n=10$

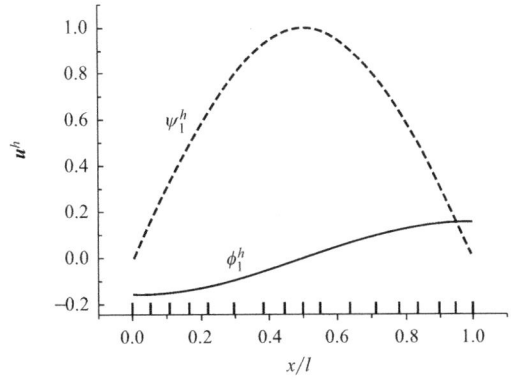

Figure 7.6   Vibration mode (ϕ_1^h and ψ_1^h) and adaptive mesh with $n=10$

7.7.2   Example 2: Moderately thick-wall thickness

The free vibration of a moderately thick cylindrical shell simply supported at both ends was considered using the following basic parameters: of the cylindrical shell are

$$r=2 \text{ m}, \ h=0.1 \text{ m}, \ l=24 \text{ m}, \ J=h^3/12,$$
$$E=200 \text{ GPa}, \ \kappa=5/6, \ \mu=0.3, \ \rho=7800 \text{ kg/m}^3 \tag{7.20}$$

The first 15 continuous eigenpairs ($k=1$-15) of the cylindrical shell free vibration under $n=0$ were computed using the proposed method, and the results are listed in Table 7.2. In paper[58], the thin shell theory was used to analyse the cylindrical shell (the transverse shear deformation of the cylindrical shell was ignored), and the dynamic stiffness method was used to solve it. The corresponding analytical solutions were obtained by Leissa[59].

Frequencies ω (rad $\cdot$ s^{-1}) of the moderately thick circular cylindrical shell   Table 7.2

k	Frequency values		
	ω_k^h	ω_k^h ①	ω_k^h ②
1	0.00002932	0.000283	0.000000
2	411.207632	411.116481	411.116481
3	660.667654	660.667696	660.667696
4	822.416495	822.232962	822.232962
5	1233.627583	1233.349444	1233.349444
6	1304.685375	1304.621327	1304.621327
7	1644.875834	1644.465925	1644.465925
8	1885.667592	1885.655130	1885.655135

continued

k	Frequency values		
	ω_k^h	$\omega_k^{h\,①}$	$\omega_k^{h\,②}$
9	2056.048683	2055.582407	2055.582407
10	2268.409842	2268.599627	2268.599665
11	2412.827456	2413.197434	2413.197434
12	2464.862857	2465.411695	2465.411695
13	2467.398542	2466.698888	2466.698888
14	2489.487637	2490.295748	2490.295748
15	2504.276532	2505.369696	2505.369696

Source: ①Results from paper [58], ②Results from paper [59].

It can be seen from Table 7.2 that the results are in good agreement with those of Chen[58] and Leissa[59] at low orders, demonstrating the applicability of the adaptive method to the continuous order special solution of cylindrical shells. It should be emphasised that with the increase in order ($k=11\text{-}15$), the difference in the frequency results becomes more significant, which is due to the error introduced by using the thin shell theory to analyse a moderately thick cylindrical shell structure.

7.7.3 Example 3: Different circumferential wave numbers and thickness-to-length ratios

The free vibrations of a moderately thick cylindrical shell simply supported at both ends under different circumferential wavenumbers n and thickness-length ratios h/l were considered. First, the free vibration of a moderately thick cylindrical shell under different circumferential ($n=1\text{-}5$) was analysed with Poisson's ratio $\mu=0.3$, length-radius ratio $l/r=1.0$, and thickness-radius ratio $h/r=0.1$. The other parameters were set as follows:

$$r=2 \text{ m}, \ J=h^3/12, \ E=200 \text{ GPa},$$
$$\kappa=5/6, \ \rho=7800 \text{ kg/m}^3 \tag{7.21}$$

The proposed method was used to solve the first-order eigenpairs under various working conditions, and the results of the natural vibration frequency were transformed into dimensionless values $\overline{\omega}=\omega l\sqrt{\rho\,(1+\mu^2)\,/E}$, as displayed in Table 7.3. Armenakas *et al.*[60] used the finite element model of the shape function of the circular displacement distribution along the cylindrical shell, whereas[61] applied the layered cylindrical shell model to solve this example and obtain effective solutions. It can be observed from Table 7.3 that the results obtained by the proposed method under different circumferential wavenumbers are in good agreement with the solutions in the literature, demonstrating that the adaptive method in this study is applicable for various circumferential wavenumbers of cylindrical shells.

Non-dimensional frequencies $\bar{\omega}$ of the moderately thick circular cylindrical shell with

different circumferential wave number n　　　　　Table 7. 3

n	Frequency values		
	$\bar{\omega}_1^h$	$\bar{\omega}_1^h$ ①	$\bar{\omega}_1^h$ ②
1	1. 06343	1. 06226	1. 06234
2	0. 88121	0. 88233	0. 88253
3	0. 80667	0. 80925	0. 80951
4	0. 89756	0. 89877	0. 89893
5	1. 11564	—	1. 12209

Source: ①Results from paper [60], ②Results from paper [61].

Furthermore, the free vibration of a moderately thick cylindrical shell under different thickness-length ratios was evaluated:

$$h/l = 0.01,\ 0.1,\ 0.2,\ 0.4,\ 0.6,\ 0.8,\ \text{and}\ 1.0 \qquad (7.22)$$

The circumferential wavenumber n was set as 0 and the thickness-radius ratio h/r were set as 0 and 0. 4, respectively. The first-order ($k=1$) eigenpairs with different h/l ratios were obtained by using the proposed method, and the results of the natural vibration frequency were transformed into dimensionless values $\bar{\omega} = (\omega h/\pi)\sqrt{\rho/G}$, as listed in Table 7. 4. As indicated, the results of this study agree well with the results estimated by Armenakas *et al.* [60] and Loy and Lam [61]. This shows that the adaptive method is applicable to cylindrical shells with different thickness-length ratios, including thin and thick walls.

Non-dimensional frequencies $\bar{\omega}$ of the moderately thick circular

cylindrical shell with different thickness-length ratios　　　　Table 7. 4

h/l	Frequency values		
	$\bar{\omega}_1^h$	$\bar{\omega}_1^h$ ①	$\bar{\omega}_1^h$ ②
0. 01	0. 01611	0. 01612	0. 01002
0. 1	0. 15291	0. 15289	0. 10000
0. 2	0. 20376	0. 20495	0. 20000
0. 4	0. 27854	0. 27540	0. 27544
0. 6	0. 42212	0. 42022	0. 42035
0. 8	0. 59775	0. 60009	0. 60033
1. 0	0. 78432	0. 79274	0. 79314

Source: ①Results from paper [60], ②Results from paper [61].

7. 7. 4　Example 4: Hinged-hinged and clamped-clamped boundary conditions

To test the applicability of the proposed method to the analysis of different boundary conditions, the free vibration of rotating cylindrical shells with typical boundary conditions, such as simply supported at both ends and fixed at both ends, was considered using Poisson's ratio $\mu=0.3$ and thickness-radius ratio $h/r=0.4$. The other parameters were:

$$r=2 \text{ m, } J=h^3/12, \ E=200 \text{ GPa, } \kappa=5/6, \ \rho=7800 \text{ kg/m}^3 \qquad (7.23)$$

The proposed method was used to solve the second-order ($k=2$) eigenpairs of cylindrical shells simply supported at both ends and fixed at both ends with different thickness-length ratios:

$$h/l=0.1, \ 0.2, \ 0.4, \ 0.6, \ 0.8, \text{ and } 1.0 \qquad (7.24)$$

The results of the natural frequencies were converted into dimensionless values $\overline{\omega}=\omega r \sqrt{(1-\mu^2)\,\rho/E}$, as presented in Table 7.5. It is evident that the results of this study are in agreement with those of the layered cylindrical shell model of Loy and Lam[61], which demonstrates the applicability of the adaptive method to cylindrical shell analysis under different boundary conditions. It is important to point out that the adaptive method in this study is universal. In addition, the error estimation and mesh partitioning techniques are independent of the boundary conditions and have the potential to be extended to complex boundary conditions such as force, elastic, and mixed boundaries in practical engineering problems.

Non-dimensional frequencies $\overline{\omega}$ of the moderately thick circular cylindrical shell with different boundary conditions   Table 7.5

h/l	Frequency values			
	Simply supported boundary (Hinged-hinged)		Fixed boundary (Clamped-clamped)	
	$\overline{\omega}_2^h$	$\overline{\omega}_2^h$ [1]	$\overline{\omega}_2^h$	$\overline{\omega}_2^h$ [1]
0.1	0.93287	0.92930	1.02652	1.03567
0.2	1.07653	1.06948	1.43737	1.47766
0.4	2.05123	2.04718	3.15265	3.27761
0.6	3.59571	3.62123	5.09947	5.32309
0.8	5.32711	5.39823	7.46493	7.39131
1.0	7.11682	7.24317	9.32662	9.46118

Source: [1]Results from paper[61].

Figures 7.7 and 7.8 respectively depict the modes of vibration under $h/l=0.1$ for the two boundary cases. The symbols $(\ )_2^h$ and $(\overline{\ })_2^h$ represent the second mode solutions under the condition of simply supported and fixed at both ends, respectively, and the final mesh distribution of the adaptive solution is marked on the horizontal axis. The mode values of u_2^h, w_2^h, ϕ_2^h, $\overline{v}_2^h$, and $\overline{\psi}_2^h$ are all 0; for simplicity, these modes are not indicated in the figure. As exhibited in Figure 7.7, the change in the second mode is relatively gentle in both domains under the condition of simply supported at both ends, and the adaptive method divides a relatively sparse and uniform mesh. Based on Figure 7.8, the second mode of vibration under the condition of fixed support at both ends changes drastically at both ends and in the middle region. The adaptive method divides relatively dense meshes, whereas relatively sparse meshes are used in other regions. The adaptive refinement optimisation under different boundary conditions shows the applicability of the adaptive method presented in this study.

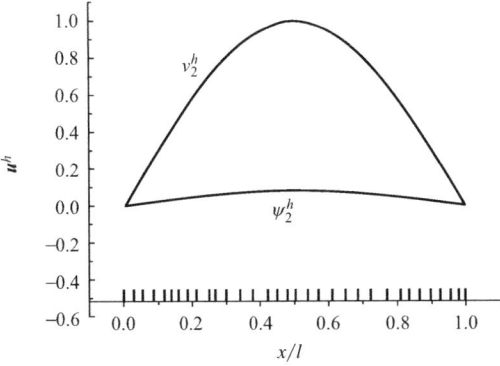

Figure 7.7　Vibration mode (v_2^h and ψ_2^h) and adaptive mesh under simply supported boundary conditions at both ends

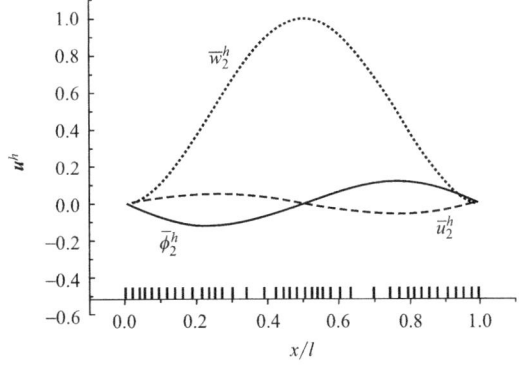

Figure 7.8　Vibration mode ($\overline{u}_2^h$, $\overline{w}_2^h$, $and\overline{\phi}_2^h$) and adaptive mesh under fixed boundary conditions at both ends

7.8　Conclusions

The main conclusions andanalyses of the above numerical examples can be summarised as follows:

(1) A superconvergent patch recovery solution for the vibration modes of rotating cylindrical shells was developed. The superconvergent patch recovery displacement method and high-order shape function interpolation techniques were introduced to obtain the superconvergent solution of the mode. Subsequently, the superconvergent solution of the mode of rotating cylindrical shells was used to estimate the errors of the finite element solutions in the energy norm, and the elements that exceeded the error limit requirements were subdivided to generate a new mesh in accordance with the errors.

(2) The finite element mesh adaptivelyrefines and optimises according to the change in vibration modes of rotating cylindrical shells, demonstrating the applicability of this method to various cases. The adaptive method can automatically divide a relatively dense finite element mesh in the region where the modal solution changes significantly, while a relatively sparse mesh is used in other regions. This method can flexibly realise the accurate computation of results by meshing according to the changes in the external conditions of the rotating cylindrical shells.

(3) Based on the analysis of the numerical examples, the proposed method is very suitable for rotating cylindrical shells with different wall thicknesses. In addition, the results of the rotating cylindrical shells by using the theory of free vibration of a moderately thick shell are more accurate and reliable than those obtained using the theory

of free vibration of a thin shell considering the effects of transverse shear deformation and moment of inertia.

(4) The proposed method is applicable to variable geometrical forms of rotating cylindrical shell structures. By analysing the various numerical examples, it is verified that in the free vibration of rotating cylindrical shells, the proposed method is reliable for solving the frequency of each order and is applicable to all types of circular wavenumber cylindrical shells and cylindrical shells with different thickness-length ratios. The applicability of this method to different boundary conditions is demonstrated under different boundary conditions, such as simple (hinged) supports and fixed (clamped) supports at both ends.

The method introduced inthis study is an adaptive method with the highest precision among all existing finite element algorithms. The results are much more accurate and reliable than those of the conventional finite element method. It can be used to study complex plate and shell problems in practical engineering, improve the current high-performance three-dimensional adaptive algorithms, and solve the frequency of solid structures, which will be the topics to be tackled in our next research.

References

[1] Ide S, Yabe S, Tanaka Y, et al. Earthquake potential revealed by tidal influence on earthquake size-frequency statistics [J]. Nature Geoscience, 2016, 9 (11): 834-837.

[2] Chestler S R, Creager K C. Evidence for a scale-limited low-frequency earthquake source process [J]. Journal of Geophysical Research: Solid Earth, 2017, 122 (4): 3099-3114.

[3] Rajesh R R, Pradeep K D, Subhamoy B, et al. Seismic behaviour of rocking bridge pier supported by elastomeric pads on pile foundation [J]. Soil Dynamics and Earthquake Engineering, 2019, 124: 98-120.

[4] Joanna M D, Radoslaw S. Simulation of dynamic behaviour of RC bridge with steel-laminated elastomeric bearings under high-energy mining tremors [J]. Key Engineering Materials, 2013, 531-532, : 662-667.

[5] Rajesh R R, Ranjan B, Bhattacharya S, et al. Application of controlled-rocking isolation with shape memory alloys for an overpass bridge [J]. Soil Dynamics and Earthquake Engineering, 2021, 149: 106827.

[6] Dey T, Ramachra. Non-linear vibration analysis of laminated composite circular cylindrical shells [J]. Composite Structures, 2017, 163: 89-100.

[7] Pang F, Tian H, Li H, et al. Semianalytical method for the free vibration characteristics analysis of cylindrical shells [J]. Journal of Vibration and Shock, 2019, 38 (22): 21-28.

[8] Bhattacharya S, Adhikari S, Alexer, et al. A simplified method for unified buckling and free vibration analysis of pile-supported structures in seismically liquefiable soils [J]. Soil Dynamics and Earthquake Engineering, 2009, 29: 1220-1235.

[9] Albdiry M T, Almensory M F. Failure analysis of drillstring in petroleum industry: A review [J]. Engineering Failure Analysis, 2016, 65: 74-85.

[10] Miklashevich I A. Delamination of composites along the interface as buckling failure of the stressed layer [J]. Mechanics of Composite Materials, 2004, 40 (4).

[11] Flores F G, Godoy L A. Forced vibrations of silos leading to buckling [J]. Journal of Sound and Vibration, 1999, 224 (3): 431-454.

[12] Fluegge S. Stresses in shells [J]. Berlin: Springer, 1973.

[13] Love A E H. A treatise on the mathematical theory of elasticity [J]. Cambridge: Cambridge University Press, 2013.

[14] Tzou H S, Howard R V. A Piezothermoelastic Thin Shell Theory Applied to Active Structures [J]. Journal of Vibration Acoustics, 1994, 116 (3): 295-302.

[15] Confalonieri F, Ghisi A, Perego U, et al. 8-Node solidshell elements selective mass scaling for explicit dynamic analysis of layered thin-walled structures [J]. Computational Mechanics, 2015, 56 (4): 585-599.

[16] Kiendl J, Bletzinger K U, Linhard J, et al. Isogeometric shell analysis with Kirchhoff-Love elements [J]. Comput Methods Appl Mech Engrg, 2009, 198: 3902-3914.

[17] Zukas J A. Effects of transverse normal and shear strains in orthotropic shells [J]. AIAA Journal, 1974, 12 (12): 1753-1755.

[18] Zareh M, Qian X. Kirchhoff-Love shell formulation based on triangular isogeometric analysis [J]. Computer Methods in Applied Mechanics and Engineering, 2018, 347: 853-873.

[19] Creaghan S G, Palazotto A N. Nonlinear large displacement and moderate rotational characteristics of composite beams incorporating transverse shear strain [J]. Computers & Structures, 1994, 51 (4): 357-371.

[20] Hansbo P, Larson M G. A posteriori error estimates for continuous/discontinuous Galerkin approximations of the Kirchhoff-Love plate [J]. Comput Methods Appl Mech Engrg, 2011, 200: 3289-3295.

[21] Repin S, Sauter S A. Estimates of the modeling error for the Kirchhoff-Love plate model [J]. C R Acad Sci Paris, 2010, 348: 1039-1043.

[22] Weise M. Residual error estimation for anisotropic Kirchhoff plates [J]. Applied Numerical Mathematics, 2018, 125: 10-22.

[23] Rychter Z. Global error estimates in reissner theory of thin elastic shells [J]. Int J Engng Sci, 1988, 26 (8): 787-795.

[24] Mohammad Z N, and Amin H. Non-local analysis of free vibration of bi-directional functionally graded Euler-Bernoulli nano-beams [J]. International Journal of Engineering Science, 2016, 105: 1-11.

[25] Lee W, Chao H. Flexural vibration analysis of a loaded double-tapered circular beam with a linearly varying wall thickness [J]. The Journal of the Acoustical Society of America, 1992, 92 (4): 2260-2263.

[26] Lin F, Xiang Y. Numerical analysis on nonlinear free vibration of carbon nanotube reinforced composite beams [J]. International Journal of Structural Stability and Dynamics, 2014, 14 (1): 1350056.

[27] Tang L, Huang D, Zhu X, et al. Forced vibration analysis of rotating ring with wave propagation method [J]. Journal of Mechanical Strength, 2013, 35 (2): 119-126.

[28] Kasper D G, Swanson D G, Reichard K M, et al. Higher-frequency wavenumber shift and frequency shift in a cracked, vibrating beam [J]. Journal of Sound and Vibration, 2008, 312: 1-18.

[29] Shu C. An efficient approach for free vibration analysis of conical shells [J]. International Journal of

Mechanical Sciences, 1996, 38 (8-9): 935-949.

[30] Lee J. Identification of multiple cracks in a beam using natural frequencies [J]. Journal of Sound and Vibration, 2009, 320 (3): 482-490.

[31] Wang Y, Ju Y, Zhuang Z, et al. Adaptive finite element analysis for damage detection of non-uniform Euler-Bernoulli beams with multiple cracks based on natural frequencies [J]. Engineering Computations, 2017, 35 (3): 1203-1229.

[32] Gerasimov T, Stein E, Wriggers P, et al. Constant-free explicit error estimator with sharp upper error bound property for adaptive FE analysis in elasticity and fracture [J]. International Journal for Numerical Methods in Engineering, 2015, 101 (2): 79-126.

[33] Zienkiewicz O C. The background of error estimation and adaptivity in finite element computations [J]. Computer Methods in Applied Mechanics and Engineering, 2006, 195 (4-6): 207-213.

[34] Bespalov A, Haberl A, Praetorius D, et al. Adaptive FEM with coarse initial mesh guarantees optimal convergence rates for compactly perturbed elliptic problems [J]. Computer Methods in Applied Mechanics and Engineering, 2017, 317: 318-340.

[35] Liu X, Wang C, Wang H, et al. Application in sheet metal forming based on h-adaptive finite element method [J]. Journal of Graphics , 2021.

[36] Arthurs C J, Bishop M J, Kay D, et al. Efficient simulation of cardiac electrical propagation using high-order finite elements II: adaptive p-version [J]. Journal of Computational Physics, 2013, 253: 443-470.

[37] Wang Y, Ju Y, Chen J, et al. Adaptive finite element-discrete element analysis for the multistage supercritical CO2 fracturing of horizontal wells in tight reservoirs considering pre-existing fractures and thermal-hydro-mechanical coupling [J]. Journal of Natural Gas Science and Engineering, 2019, 61: 251-269.

[38] Wang Y. An h-version adaptive FEM for eigenproblems in system of second order ODEs: Vector Sturm-Liouville problems and free vibration of curved beam [J]. Engineering Computations, 2020, 37 (1): 1210-1225.

[39] Gomez-Revuelto I, Garcia-Castillo L E, Llorente-Romano S, et al. A three-dimensional selfadaptivehpfinite element method for the characterization of waveguide discontinuities [J]. Computer Methods in Applied Mechanics and Engineering, 2012, 249: 62-74.

[40] Wang Y, Wang J. An hp-version adaptive finite element algorithm for eigensolutions of free vibration of moderately thick circular cylindrical shells via error homogenization and higher-order interpolation [J]. Engineering Computations, 2021.

[41] Arndt M, Machado R D, Scremin A, et al. An adaptive generalized finite element method applied to free vibration analysis of straight bars and trusses [J]. Journal of Sound and Vibration, 2010, 329 (6): 659-672.

[42] Arndt M, Machado R D, Scremin A, et al. Accurate assessment of natural frequencies for uniform and non-uniform Euler-Bernoulli beams and frames by adaptive generalizedfinite element method [J]. Engineering Computations, 2016, 33 (5): 1586-1609.

[43] Stein E, Seifert B, Ohnimus S, et al. Adaptive finite element analysis of geometrically non-linear plates and shells, especially buckling [J]. International Journal for Numerical Methods in Engineering, 1994, 37 (15): 2631-2655.

[44] Ma L, Chang J. An iterative adaptive median filter algorithm [J]. Computer Engineering & Software, 2020, 41 (9): 69-71.

[45] Civalek O, Akgoz B. Free vibration analysis of microtubules as cytoskeleton components: nonlocal Euler-Bernoulli beam modeling [J]. Scientia Iranica, 2010, 17 (5): 367-375.

[46] Wang X, Yuan Z. Techniques for vibration analysis of hybrid beam and ring structures with variable thickness [J]. Computers and Structures, 2018, 206: 109-121.

[47] Yuan S, Wu Y, Xing Q, et al. Recursive super-convergence computation for multi-dimensional problems via one-dimensional element energy projection technique [J]. Applied Mathematics and Mechanics, 2018, 39 (7): 1031-1044.

[48] Krause R, Rank E. Multiscale computations with a combination of the h-and p-versions of the finite-element method [J]. Computer Methods in Applied Mechanics and Engineering, 2003, 192: 3959-3983.

[49] Noor A K, Burton W S. Stress and free vibration analyses of multilayered composite plates [J]. Composite Structures, 1989, 11: 183-204.

[50] Hou L, Peters D A. Application of triangular space-time finite elements to problems of wave propagation [J]. Journal of sound and vibration, 1994, 173: 611-632.

[51] Huang H, Huang Y. Displacemental vibration equations of thick shells of revolution considering transverse shear deformation [J]. Spatial Structures, 2003, 10 (1): 3-7 (in Chinese).

[52] Clough R W, Penzien J. Dynamics of structures [J]. New York: McGraw-Hill, 1993.

[53] Wiberg N E, Bausys R, Hager P, et al. Adaptive h-version eigenfrequency analysis [J]. Computers and Structures, 1999, 71 (5): 565-584.

[54] Wiberg N E, Bausys R, Hager P, et al. Improved eigenfrequencies and eigenmodes in free vibration analysis [J]. Computers and Structures, 1999, 73 (1-5): 79-89.

[55] Zienkiewicz O C, Zhu J. The superconvergent patch recovery and a posteriori error estimates. Part 2: error estimates and adaptivity [J]. International Journal for Numerical Methods in Engineering, 1992, 33 (7): 1365-1382.

[56] Sivadas K R, Ganesan N. Free vibration and material damping analysis of moderately thick circular cylindrical shells [J]. Journal of Sound and Vibration, 1994, 172 (1): 47-61.

[57] Chen X, Ye K. Analysis of free vibration of moderately thick circular cylindrical shells using the dynamic stiffness method [J]. Engineering Mechanics, 2016, 33 (9): 40-48.

[58] Chen X. Dynamic Stiffness Method for Free Vibration of Rotating Shell [J]. Beijing: Tsinghua University, 2009.

[59] Leissa A W. Vibration of shells [J]. Washington D C: NASA. , 1973.

[60] Armenakas A E, Gazis D S, Herrmann G, et al. Free vibrations of circular cylindrical shells [J]. Oxford: Pergamon Press, 1969.

[61] Loy C T, Lam K Y. Vibration of thick cylindrical shells on the basis of three-dimensional theory of elasticity [J]. Journal of Sound and Vibration, 1999, 226 (4): 719-737.

Chapter 8
Improved *hp*-version adaptive finite element method for vibration of cylindrical shells

8. 1   Introduction

Cylindrical shells are widely used invarious engineering applications including civil, mechanical, aerospace, and marine engineering owing to their symmetrical structure and simple manufacturing process. Studying and analysing the natural frequency, mode shape, and other dynamic structural characteristic parameters is essential for ensuring a reliable, safe, and efficient performance[1]. Shells are categorised according to their radius-to-thickness ratio, as defined in Table 8. 1. To date, research into shell vibration has focused predominantly on thin shells, based on the classical theory of thin shells[2-5] while ignoring transverse shear deformations[5, 6]. However, practical engineering situations often involve thicker plate and shell structures that lie beyond the application scope of this theory. Such situations make it necessary to consider the influence of transverse shear deformations and moments of inertia on vibration. This has motivated further research into models of moderately thick circular cylindrical shells. Shear deformations associated with the wall thickness are considered in a medium-thickness cylindrical shell, but geometric features (e. g. , wall thickness, radius, and length)[7-11], are key factors affecting the dynamic shell characteristics, which challenges conventional dynamic analysis methods.

Shell categories defined according to the radius-to-thickness ratio R/h Table 8. 1

Radius-to-thickness ratio	$R/h \geqslant 20$	$6 < R/h < 20$	$R/h \leqslant 6$
Category	Thin	Moderately thick	Thick

Analytical or semi-analytical methods and models have beenused to study vibrational characteristics in modal analyses of cylindrical shell structures of variable thickness. Modal analysis includes the study of the natural frequency and vibrational modes. Various studies have computed the frequency and mode shape of a structure[12-14] and studied the free vibration of thin circular cylindrical shells using thin-shell theory. The theory, based on the Kirchhoff-Love hypothesis[5] and the wave propagation method[15-17], establishes the differential equations applicable to thin shells. However, these equations are ill-suited to moderately thick cylindrical shells. Thin-shell theory, used to study cylindrical shell vibrations, is accurate only when analysing long shells under simply supported boundary conditions. The semi-analytical method[8], the dynamic stiffness method[18], and the symplectic approach[19, 20] are commonly used to study the free vibration of cylindrical shells, but they are poorly applicable to complex geometrical structures. Ahad et al.[9] used the improved Fourier-series method to study the free vibration of isotropic homogeneous moderately thick open cylindrical shells with arbitrary opposite rotation angles and under general elastic constraints. Shi et al.[10] used a semi-analytical method to analyse the free vibration characteristics of moderately thick composite laminated cylindrical shells under arbitrary classical and elastic boundary conditions. Chen and Ye[21] used the dynamic stiffness method to study the free vibration of moderately thick and thin cylindrical shells. The layer-wise method[22, 23] and Reissner-mixed variational theorem[24, 25] are suitable for reliable solutions of the vibration modes and frequencies of heterogeneous and multilayered structural materials, such as thin/thick laminated structures, laminated composites, functionally gradient materials. The high-precision eigensolutions of free vibration of moderately thick circular cylindrical shells have been studied by taking into account the variable geometrical factors. This was done by studying the effects of circumferential wave numbers on the dynamic solutions using the Flügge equations of motion[26], theory and experiments[27], the three-dimensional theory of elasticity[28], and refined higher-order theory[29]. The thickness-to-radius or -length ratio was investigated using three-dimensional linear elasticity and an iterative approach[30], the spline function method[31], the improved Fourier-series method[9], Hamilton's principle and first-order shear deformation theory[10], and the dynamic stiffness method[11].

Practical numerical models and methods have been developed to solve all types of cylindrical shells. In particular, the finite element method (FEM) is widely used to perform free vibration analysis for complex structural members of thin circular cylindrical shells[32]. The conventional FEM is highly versatile and applicable to any structural configuration. However, the density of the designed mesh and the computational cost grow with the required solution accuracy. Moreover, the accuracy of the FEM as a whole depends on the particular meshing method employed, as well as on user experience. In contrast, the adaptive FEM automatically adjusts the algorithm to seek the optimal solution to achieve the targeted accuracy[33, 34], reduce the computation cost, improve the process of conventional finite element (FE) mesh design, and improve the efficiency and accuracy of the solution[35, 36]. It mainly includes the *h*-version adaptive method for increasing the mesh density[37, 38], the *p*-version adaptive method for increasing the element order[39, 40], and the *hp*-version adaptive method[41-43]. The adaptive FEM shows good potential for producing high-precision solutions to structural eigenvalue problems, e. g., involving the free vibration and elastic stability of beams, plates, and shells[44-46]. The *h*-version adaptive FEM yields high-precision characteristic solutions, but its accuracy depends on the mesh density. A low-order mesh is too dense and the convergence speed is slow. The *p*-version adaptive FEM converges fast but the disadvantage of mixing high and low orders results in local node redundancy. This redundancy is also encountered in the *hp*-version adaptive FEM. Building upon previous research into the *h*-version adaptive FEM[47, 48], the present study providesa reasonable introduction todisplacement superconvergent patch recovery, mesh refinement, and higher-order interpolation techniques. A novel *hp*-version adaptive FEM is proposed for investigating the high-precision eigensolutions of the free vibration of moderately thick circular cylindrical shells. The approach considers variable geometrical factors, such as the thickness, circumferential wave number, radius, and length.

The remainder of this chapter is structured as follows. Section 8.2 introduces differential equations for the free vibration of moderately thick circular cylindrical shells. Section 8.3 then introduces FE discretisation and the solutions of the differential equations. Section 8.4 presents the hp-version adaptive FEM via error homogenisation and higher-order interpolation. This method combines the h-version mesh refinement and p-version higher-order interpolation. Section 8.5 discusses the global algorithm and procedure. Section 8.6 provides representative numerical examples illustrating the performance of the proposed method. Finally, Section 8.7 summarises the main conclusions.

8.2　Differential equations describing the free vibration of moderately thick circular cylindrical shells

The geometrical model, coordinate systems, and parameters describing the moderately thick circular cylindrical shell investigated in this study are shown in Figure 8.1. The rotation axis is denoted ox. The local coordinate system at a point A on the mid-plane is defined by $\alpha\beta\gamma$, where α points along to the tangential direction of the meridian (the direction of the shell axis), and β points along the tangential direction of the weft circle, and γ points along the normal direction. The five independent displacements of the shell are the linear displacements u, v, and w along the α, β, and γ directions, and the angular displacements ϕ and ψ defined around the γ and β directions, respectively. The radius of the middle shell surface is r, the section thickness is h, the shear stiffness correction factor of the section is κ, the moment of inertia is J, the length is l, the elastic modulus of the material is E, the shear modulus is G, Poisson's ratio is μ, and the density is ρ.

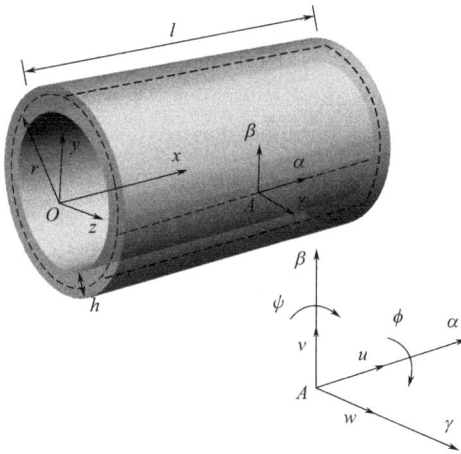

Figure 8.1　Geometrical model, coordinate systems, and symbols describing a moderately thick circular cylindrical shell

The governing differential equations of the free vibration of the shell are:

$$(1/2)K\left[(1-\mu)n^2 u/r^2 - 2u'' - (1+\mu)nv'/r - 2\mu w'/r\right]$$
$$= \omega^2 \rho h u \tag{8.1a}$$

$$(1/2)K\left[(1+\mu)nv'/r + 2n^2 v/r^2 - (1-\mu)(1+h^2/12r^2)v'' + 2nw/r^2\right.$$
$$\left. + (1-\mu)nh^2\phi'/(12r^2) - (1-\mu)h^2\psi''/(12r)\right] \tag{8.1b}$$
$$+ \bar{\kappa}Gh(v/r^2 + nw/r^2 - \psi/r) = \omega^2 \rho h v$$

$$K(\mu u'/r+nv/r^2+w/r)+\bar{\kappa}Gh(n^2w/r^2+nv/r^2-n\psi/r-w''-\phi')$$

$$=\omega^2\rho hw \tag{8.1c}$$

$$(h^2/24)K[-(1-\mu)nv'/r^2+(1-\mu)n^2\varphi/r^2-2\phi''-(1+\mu)n\psi'/r]$$

$$+\kappa Gh(w'+\phi)=\omega^2\rho J\phi \tag{8.1d}$$

$$(h^2/24)K[-(1-\mu)v''/r+(1+\mu)n\phi'/r+2n^2\psi/r^2-(1-\mu)\psi'']$$

$$+\kappa Gh(\psi-v/r-nw/r)=\omega^2\rho J\psi \tag{8.1e}$$

where the prime mark ($'$) denotes the derivative with respect to the independent variable x, $K=Eh/(1-\mu^2)$ is the shell stiffness, ω is the frequency, and n is the circumferential wave number.

The eigenproblem solved in this study consists of finding the eigenvalues λ and the associated n_d dimensional vector eigenfunctions $\boldsymbol{u}(x)=(u_1(x),\cdots,u_{n_d}(x))^{\mathrm{T}}$ for the following system of second-order ordinary differential equations (ODEs) (Greenberg, 1991; Kurochkin, 2014):

$$\boldsymbol{Lu}\equiv\boldsymbol{Au''}+\boldsymbol{Bu'}+\boldsymbol{Cu}=\lambda\boldsymbol{Ru}, \quad a<x<b, \tag{8.2}$$

where $\boldsymbol{L}$ is the differential operator, and $\boldsymbol{A}$, $\boldsymbol{B}$, $\boldsymbol{C}$, and $\boldsymbol{R}$ are continuous $n_d\times n_d$ matrix functions on (a,b), and $n_d=5$ is associated with the governing differential equations for the free vibration of the shell, as shown in Equation (8.1). Structural vibration problems were chosen to illustrate the possible physical interpretations of the equations. The corresponding structural natural frequencies ($\omega=\sqrt{\lambda}$) and modes represent the eigenvalues and vector eigenfunctions, respectively.

8.3　Finite element discretisation and solutions of the differential equations

8.3.1　Higher-order Lagrange shape functions

The Lagrange interpolation formula provides a simple construct for higher-order shape functions that satisfy Equation (8.2):

$$l_a^p(\xi)=\prod_{\substack{b=1\\b\neq a}}^{n}\frac{(\xi-\xi_b)}{(\xi_a-\xi_b)}=$$

$$\frac{(\xi-\xi_1)(\xi-\xi_2)(\cdots)(\xi-\xi_{a-1})(\xi-\xi_{a+1})(\cdots)(\xi-\xi_n)}{(\xi_a-\xi_1)(\xi_a-\xi_2)(\cdots)(\xi_a-\xi_{a-1})(\xi_a-\xi_{a+1})(\cdots)(\xi_a-\xi_n)} \tag{8.3}$$

The order of this polynomial is $p=n-1$. Having chosen the end-node locations, the internal values of ξ_a may be spaced in uniform increments. For one-dimensional elements, we can set[49]:

$$N_a(\xi)=l_a^p(\xi) \tag{8.4}$$

to define the shape functions.

8.3.2　Finite element discretisation

The weak form for the eigenproblems in the system of second-order ODEs, as defined in Equation (8.2), can be expressed as:

$$\int_{\Omega} \{ v'^{\mathrm{T}}Au' + v^{\mathrm{T}}[Bu' + (C - \lambda R)u] \} dx = 0 \tag{8.5}$$

where v is a trial function and Ω is the solution domain. The FE model uses the conventional degree m for the polynomial elements. We let e denote a typical element with end-node coordinates $\overline{x}_1$ and $\overline{x}_2$ and with length h. We write the trial function on an element of degree m as:

$$v = \sum_{i=1}^{m+1} N_i v_i \tag{8.6}$$

where N_i is an $n_d \times n_d$ shape function matrix defined by:

$$N_i = N_i I, \ i = 1, 2, \cdots, m+1 \tag{8.7}$$

where I is the $n_d \times n_d$ identity matrix.

Using the conventional FEM, the element stiffness and mass matrices (K^e and M^e, respectively) are computed and assembled to form the global stiffness and mass matrices K and M. The FE equation is then derived as an eigenvalue equation in the following matrix form:

$$KD = \lambda MD \tag{8.8}$$

where D is the eigenfunction vector, and matrices K and M are independent of λ. Given an arbitrary trial value λ_a as the shift value, Equation (8.8) is equivalently written in the shifted form[49]

$$K_a D = \mu MD \text{ with } K_a = K - \lambda_a M, \ \mu = \lambda - \lambda_a \tag{8.9}$$

In the proposed method, theconventional FE computation for eigenpair solutions is based on the Sturm sequence property[50, 51], which can be expressed as:

$$K - \omega^2 M = LD(\omega)L^{\mathrm{T}} \tag{8.10}$$

where L is a lower triangular matrix with leading diagonal elements equal to unity, L^{T} is its transpose, and $D(\omega)$ is a diagonal matrix.

Based on the resultfor D, the following Rayleigh quotient[50, 51] is used to estimate the eigenvalue solution to accelerate its convergence:

$$\lambda = \frac{D^{\mathrm{T}}KD}{D^{\mathrm{T}}MD} \tag{8.11}$$

8.3.3 Inverse iteration technique for eigensolutions

The Sturm theorem[50-54] is introduced to fix the eigen intervals, so that each interval contains a single eigenvalue or coincident eigenvalues. This technique was a widely used to evaluate the upper or lower bounds of eigenvalues of any order. Consequently, using the conventional FEM, this research applies the inverse iteration to solve for single eigensolutions and applies subspace iteration to solve for multiple eigensolutions. The inverse iteration technique is used to solve for the single eigenpairs as follows:

$$\left.\begin{array}{l}\overline{\boldsymbol{D}}_{i+1}=\boldsymbol{K}_a^{-1}\boldsymbol{M}\boldsymbol{D}_i \\[2mm] \mu_{i+1}=\dfrac{\overline{\boldsymbol{D}}_{i+1}^{\mathrm{T}}\boldsymbol{M}\boldsymbol{D}_i}{\overline{\boldsymbol{D}}_{i+1}^{\mathrm{T}}\boldsymbol{M}\overline{\boldsymbol{D}}_{i+1}} \\[3mm] \boldsymbol{D}_{i+1}=\mathrm{sgn}(\mu_{i+1})\dfrac{\overline{\boldsymbol{D}}_{i+1}}{\max(\overline{\boldsymbol{D}}_{i+1})}\end{array}\right\}\quad i=0,\ 1,\ \cdots \qquad (8.12)$$

where i is the loop index.

The above inverse iterationprocedure is terminated when the following condition is met:

$$|\mu_{i+1}-\mu_i|<Tol \quad \text{and} \quad \max|D_{i+1}-D_i|<Tol \qquad (8.13)$$

8.4　*hp*-version adaptive finite element method via error homogenisation and higher-order interpolation

8.4.1　*h*-version mesh refinement

The superconvergent patch recovery displacement method was developed[47, 55, 56] to acquire the superconvergent displacements of the FE solutions in static and dynamic problems. The displacements provided by this method can be applied to eigenfunctions. For example, if element e is a superconvergent computation element and elements $e-1$ and $e+1$ can be considered its neighbouring elements, all FE nodes in patched elements $e-1$, e, and $e+1$ are selected for the computation process. Further, the superconvergent displacements for element e can be computed as[47]:

$$\boldsymbol{u}^*(x)=\sum_{i=1}^{r}\boldsymbol{N}_i(x)\boldsymbol{u}_i^h+\sum_{i=1}^{s}\boldsymbol{N}_i(x)\overline{\boldsymbol{u}}_i^* \qquad (8.14)$$

where $r=2$ is the number of end nodes, s is the number of internal nodes, and $\boldsymbol{N}_i$ is the shape function matrix. Using high-order shape function interpolation, the polynomial order of the shape function is increased, $r+s>m+1$. To optimise the superconvergent order $O(h^{2m})$ for displacements at the end nodes, the displacement recovery field can be expressed for the FE nodes as:

$$\overline{u}_i^*(x)=\boldsymbol{Pa},\ i=1,\ \cdots,\ n_d \qquad (8.15)$$

where $\boldsymbol{P}$ is the given function vector and $\boldsymbol{a}$ can be determined by least-squares fitting for the coincidence of displacements at the end nodes in both the recovery and conventional FE fields. The superconvergent displacements of the recovery field are used in Equation (8.14) to obtain the superconvergent solutions of element e. We use the following forms for the vector coefficients $\boldsymbol{P}$ and $\boldsymbol{a}$:

$$\boldsymbol{P}=\begin{bmatrix}1 & x & \cdots & x^p\end{bmatrix},\ \boldsymbol{a}=\begin{bmatrix}a_1 & a_2 & \cdots & a_m\end{bmatrix}^{\mathrm{T}} \qquad (8.16)$$

The value of $\boldsymbol{a}$ was determined from the minimum value of the following functional, so that the product of $\boldsymbol{P}$ and $\boldsymbol{a}$, calculated using Equation (8.16), matches the displacement values:

$$\Pi = \sum_{j=1}^{n} (u_i^*(x_j) - \mathbf{P}(x_j)\mathbf{a}), \ i = 1, \ \cdots, \ n_d \tag{8.17}$$

where n is the node number of all elements patched together.

The least-squares method applied to Equation (8.18) yields:

$$\mathbf{a} = \mathbf{A}^{-1}\mathbf{b} \tag{8.18}$$

with the coefficient matrices $\mathbf{A}$ and $\mathbf{b}$ defined as:

$$\mathbf{A} = \sum_{j=1}^{n} \mathbf{P}(x_j)^{\mathrm{T}}\mathbf{P}(x_j), \ \mathbf{b} = \sum_{j=1}^{n} \mathbf{P}(x_j)^{\mathrm{T}}u_i^h(x_j), \ i = 1, \ \cdots, \ n_d \tag{8.19}$$

After $\mathbf{a}$ is determined, the superconvergent solutions of the displacement of the piecewise elements is obtained from Equation (8.14). The estimated eigenvalue has a stationary value when all the possible functions satisfying the essential boundary conditions are considered. These stationary values are the superconvergent eigenvalues. Further, the superconvergent solutions of the displacements can be used in the Rayleigh quotient[50, 51] to estimate the eigenvalue as:

$$\lambda^* = \frac{a(\mathbf{u}^*, \ \mathbf{u}^*)}{b(\mathbf{u}^*, \ \mathbf{u}^*)} \tag{8.20}$$

where $a(\bullet)$ and $b(\bullet)$ are the strain and kinematic energy inner products, respectively. To determine whether the solution for the mesh considered meets the required tolerance condition, the error had to satisfy the following condition:

$$\|\mathbf{e}^*\| \leqslant Tol \cdot [(\|\mathbf{u}^h\|^2 + \|\mathbf{e}^*\|^2)/n_e]^{1/2} \tag{8.21}$$

where n_e is the number of elements, and $\|\mathbf{e}^*\|$ is the error in the energy norm:

$$\|\mathbf{e}^*\| = [a(\mathbf{e}^*, \ \mathbf{e}^*)]^{1/2} = \left[\int_{\Omega} \mathbf{e}^{*\mathrm{T}}\mathbf{L}\mathbf{e}^* \, \mathrm{d}x\right]^{1/2} \tag{8.22}$$

where $\mathbf{e}^* = \mathbf{u}^* - \mathbf{u}^h$ and Ω is the solution domain. Equation (8.21) can be rewritten as

$$\xi \leqslant 1 \tag{8.23}$$

where

$$\xi = \frac{\|\mathbf{e}^*\|}{\bar{e}} \quad \text{with} \quad \bar{e} = Tol \cdot [(\|\mathbf{u}^h\|^2 + \|\mathbf{e}^*\|^2)/n_e]^{1/2} \tag{8.24}$$

If Equation (8.23) is not satisfied, the corresponding element is subdivided into identical sub-elements by inserting interior nodes through h-refinement[47]. These are computed using:

$$h_{\text{new}} = \xi^{-1/m}h_{\text{old}} \tag{8.25}$$

where h_{new} is the length of the sub-element and h_{old} is the original length of element e. The above element subdivision approach is implemented as:

$$n_{\text{new}} = \min(\lfloor \xi^{1/m} \rfloor, \ d) \tag{8.26}$$

where n_{new} is the number of sub-elements after element subdivision, the symbol $\lfloor \bullet \rfloor$ denotes the "floor" operator (i. e. , the rounding down to the nearest integer), and d is the limit needed to avoid too many redundant elements. Each element e that does not meet the pre-specified error tolerance threshold is uniformly subdivided by h-version mesh refinement.

Anon-uniform mesh in a high-performance adaptive analysis process is a crucial factor for efficient computation, where a minimal number of elements and an optimised node distribution rapidly reduce the solution errors and yield high-precision solutions[44, 57, 58]. In the present study, we estimated the solution error of one element, based on the error estimation in the energy norm, using the superconvergent solution for each element. The error may be accurately reduced through a uniform subdivision for this element. Meanwhile, to avoid redundant elements in such a uniform subdivision, the proposed method limits the maximum number of subdivisions at each adaptive stage. The mesh in the global domain is non-uniform because each element is independent of the error estimation and the mesh subdivision. As the number of elements increases through subdivision, the non-uniformity of the elements reflects the variation of the eigenfunctions. The error estimation in the energy norm and uniform subdivision of each element for mesh refinement demonstrate the accuracy, reliability, and effectiveness of the eigensolutions. This is demonstrated in the numerical examples discussed below in this chapter, and in other verification examples discussed in previous studies[47].

8.4.2 p-version higher-order interpolation

After the global error ismade uniform, the order of each element is increased by the *p*-version adaptive FEM. This reduces the global error and improves the accuracy of computation results:

$$v = \sum_{i=1}^{\overline{m}+1} N_{i+1} v_{i+1} \tag{8.27}$$

where $\overline{m}$ ($\overline{m} > m$) is the number of new elements and N_{i+1} is an $n_d \times n_d$ shape function matrix defined by:

$$N_{i+1} = N_{i+1} I, \ i = 1, 2, \cdots, \overline{m}+1 \tag{8.28}$$

8.5 Global algorithm and procedure

Consider the situation thatrequires solving for a pair (ω, $\pmb{u}$) of $\pmb{L}\pmb{u} = \omega^2 \pmb{R}\pmb{u}$ for the free vibration of moderately thick circular cylindrical shells. The goal of the adaptive solution in this study is as follows. In the case where an accurate solution (ω, $\pmb{u}$) is unknown, and the error tolerance *Tol* is specified in advance, we seek an optimised finite element mesh π, such that the FE solution on the mesh (ω^h, $\pmb{u}^h$) also satisfies:

$$|\omega - \omega^h| \leqslant \text{Tol} \cdot (1 + |\omega|) \tag{8.29}$$

$$\max_{a < x < b} |u_i(x) - u_i^h(x)| \leqslant Tol, \ \max_i (\max_{a < x < b} |u_i(x)|) = 1, \ i = 1, 2, \cdots, 5 \tag{8.30}$$

In practice, because the exact solution (ω, $\pmb{u}$) is unknown in advance, the above target cannot be used as a shutdown criterion. Because the mode shape (displacement) superconvergent patch recovery solution $\pmb{u}^*$ has a higher convergence order than $\pmb{u}^h$ (i.e.,

u^* is nearer the exact solution u than u^h), u^* therefore replaces u to estimate and control the error of u^h. This yields an error estimation in the form of an energy mode. The displacement superconvergent solution is then used to calculate the Rayleigh quotient[59] and to obtain the superconvergent solution for the frequency. The frequency superconvergent solution thus obtained has a higher convergence order than the mode-shaped superconvergent solution. In this study, by estimating and controlling the error of the mode-shape solution, the frequency solution could be guaranteed to meet the error requirements. Thus, the cost method can control the mode error using the shutdown criterion.

The flowchart of the hp-version adaptive FE algorithm and the procedure for determining the eigensolutions of free vibration of a moderately thick circular cylindrical shell are shown in Figure 8.2. In h-version mesh refinement stage, based on the error estimation of the FE solution, the mesh is refined to make the global error uniform. In p-version enrichment stage, using this refined mesh, by increasing the order of the shape function to reduce the error tolerance, the FE solution and error estimation are repeated. If the solution does not meet the new error tolerance threshold, the element order increases again. Then, the optimised mesh and high-precision frequency and mode solutions that meet the error tolerance condition can be obtained, forming a complete set of hp-version adaptive FE schemes.

The procedure involves the following four basic techniques:

(1) **FE solution:** Conventional FE computation is performed based on the initial mesh. The single feature pair is iterated inversely, and the overlapped feature pair is iterated in the subspace to obtain the FE solution under the current mesh (ω^h, u^h).

(2) **Error estimation:** Based on the FE solution (ω^h, u^h), the superconvergent solution u^* is obtained using the superconvergent algorithm of FE post-processing, and the superconvergent solution of frequency ω^* is computed from the Rayleigh quotient. At the same time, the error estimates between the superconvergent and FE solutions can be obtained.

(3) **h-version mesh refinement using error estimation and element subdivision:** For the element whose error estimation does not meet the error tolerance condition, the mesh subdivided refinement method is used to subdivide it to obtain a new FE mesh, before returning to step (1). If all the elements meet the error tolerance condition, no subdivision is necessary, and the solution process is completed.

(4) **p-version enrichment using higher-order shape functions for interpolation:** Under the optimised mesh obtained in the above steps, the error tolerance is reduced, the element order increased, the FE computation continues, and the new error tolerance is used to estimate the error. For the element whose error estimate does not meet the requirements, this step is repeated, and the order of the element is increased until the new specified error tolerance condition is met, completing the solution process.

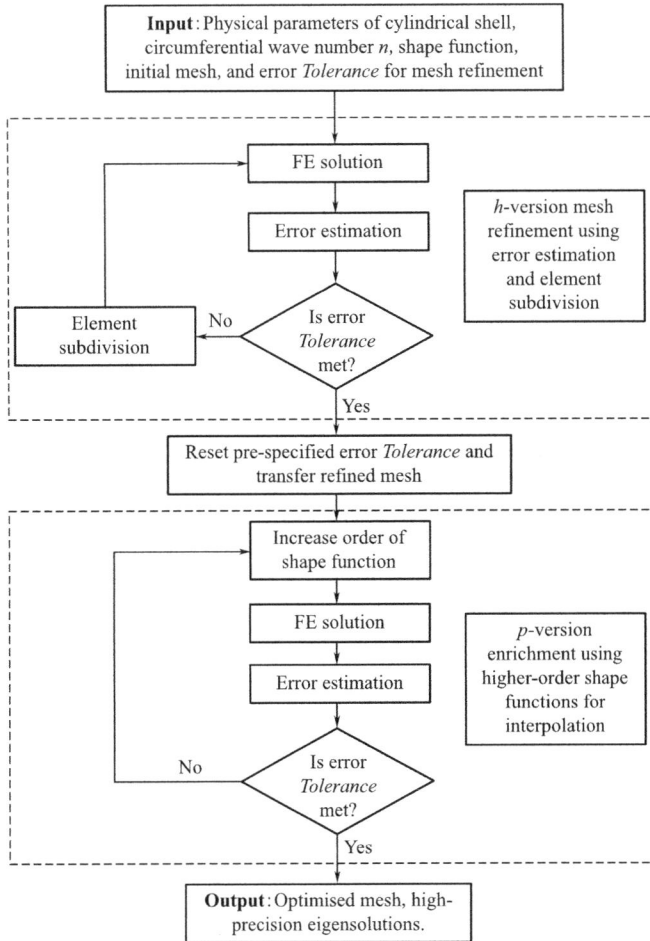

Figure 8. 2   Flow chart of the *hp*-version adaptive finite element algorithm and procedure for finding the eigensolutions of the free vibration of a moderately thick circular cylindrical shell

8. 6   Numerical examples

This sectionprovides some representative numerical examples to demonstrate the performance of the proposed method and algorithm, which were implemented in Fortran 90. The program was run on a DELL Optiplex 380 Intel (R) Core (TM) 2. 93 GHz desktop computer. Typical and representative numerical examples are provided to solve the free vibration of moderately thick cylindrical shells by using the *hp*-version adaptive FEM. The efficiency of the method was demonstrated by comparing the results of the *hp*- and *h*-version adaptive FEM, and its accuracy was verified by comparing the results for a typical moderately thick shell and a thin shell. Subsequently, the computed solutions of the free vibration frequency and the mode shape of a moderately thick cylindrical shell with different circumferential wave numbers, thickness-to-radius and -length ratios were analysed to prove the versatility and reliability of the method. Table 8. 2 lists the basic computational parameters used in the proposed adaptive FE procedure. The degree of element for mesh refinement is $m=3$; the initial number of element used in the computation procedure is $n=2$; the first order ($k=1$) of eigensolutions are solved; the tolerance for mesh refinement stage is $Tol_h=10^{-5}$; and the stricter pre-specified error tolerance for p-version enrichment stage is $Tol=10^{-6}$.

Basic computational parameters used in the adaptive finite element procedure   Table 8. 2

Parameter	Value
Degree m of element for mesh refinement	3
Initial number of element n	2
Order of eigensolution k	1
Tolerance for mesh refinement Tol_h	10^{-5}
Pre-specified error tolerance Tol	10^{-6}

The relativeerror ε_ω between the exact frequency $\overline{\omega}$ (i. e. , the high-precision solutions from other methods) and the computed frequency ω^h was defined and used to analyse the precision of the solutions:

$$\varepsilon_\omega=\frac{|\overline{\omega}-\omega^h|}{1+|\overline{\omega}|} \qquad (8. 31)$$

To analyse the vibration mode solutions, we derived the spatial distribution of the mode components and the corresponding adaptive non-uniform mesh distribution.

8. 6. 1   Example 1: h-version and hp-version adaptive finite element analysis

To verify the efficiency of the proposed algorithm, the results of the present method were compared with those of the established h-version adaptive FEM[47, 48]. Consider the free vibration of a moderately thick cylindrical shell simply supported at both ends and characterised by the following basic parameters:

$$r=148. 234 \text{ mm}, \ h=0. 508 \text{ mm}, \ l=298. 2 \text{ mm}, \ J=h^3/12,$$
$$E=203. 5 \text{ GPa}, \ \kappa=5/6, \ \mu=0. 285, \ \rho=7846 \text{ kg/m}^3 \qquad (8. 32)$$

The computed results for frequencies (order of eigensolution $k=1$) utilising the h- and hp-version adaptive FE algorithms under identical pre-specified error tolerance levels $Tol=10^{-6}$ are compared in Table 8. 3. For wave numbers $n=1$, 2, and 3, the errors between the first-order frequency values computed by the two methods fully meet the error limitation requirements. In addition, the number of elements used in the h-version method were 60, 56, and 60, whereas fewer elements (30, 30, and 28, respectively) were used with the hp-version method, thereby reducing the number of elements by nearly half. This table shows that the hp-version adaptive method achieves almost the same accuracy for the natural frequency only under half of the meshes, as compared to the h-version adaptive method.

154

Comparison of computed results for the frequencies of Example 1, utilizing the
h-[47, 48] and hp-version adaptive finite element algorithms   Table 8.3

Wave number	Frequencies		Errors		Final number of elements	
n	ω_h^h	ω_{hp}^h	$\varepsilon_{\omega h}$	$\varepsilon_{\omega hp}$	n_h	n_{hp}
1	3270.954744	3270.954633	3.41×10^{-8}	3.41×10^{-8}	60	30
2	1862.175448	1862.175355	4.95×10^{-8}	4.95×10^{-8}	56	30
3	1101.865104	1101.865039	5.95×10^{-8}	5.95×10^{-8}	60	28

Figure 8.3 shows the vibration mode diagram of the five mode components along the axis of the shell of the hp-version adaptive algorithm and the final adaptive mesh. For convenience, the mode components (u_1^h, v_1^h, w_1^h) with smaller linear displacements are shown in Figure 8.3 (a), and the mode components (ϕ_1^h, ψ_1^h) with larger angular displacements in Figure 8.3 (b). The node distribution of the element end in the entire region is shown as tick marks on the horizontal x-axis. The vibration mode diagram displays some relatively significant changes in the middle region. The adaptive method divides the non-uniform mesh, whereas the two end regions are relatively sparse. In order to obtain the same accuracy, the conventional FEM adopts uniform dense mesh, which will take more computation duration than the adaptive solution process.

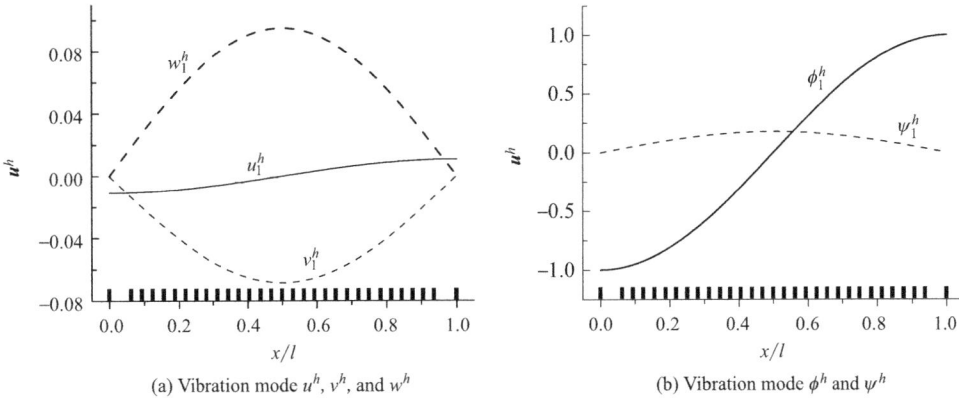

(a) Vibration mode u^h, v^h, and w^h (b) Vibration mode ϕ^h and ψ^h

Figure 8.3   Example 1: computed vibration modes along the shell axis and corresponding final meshes ($n=1$, hp-version adaptive FEM)

The errors in the computed frequencies and corresponding final meshes, utilising the h-[47, 48] and hp-version adaptive FE algorithms are shown in Figure 8.4 (with an identical pre-specified error tolerance level $Tol = 10^{-6}$). The left and right axes represent the frequency and error values, respectively. The frequency results for the h- and hp-version algorithms are clearly consistent. Also, the errors between the two algorithms increase gradually with the circumferential wave number. The lower horizongtal x-axis represents the densely distributed final mesh for the circumferential wave number $n=1$ of the h-

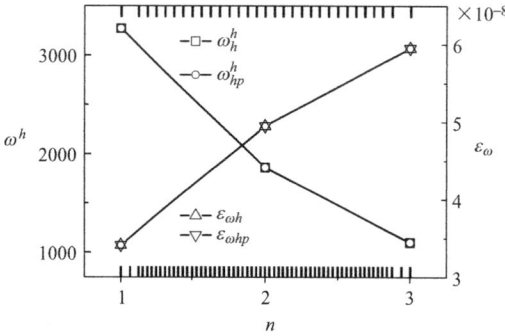

Figure 8.4 Example 1: errors in the computed frequencies and corresponding final meshes, between the h-[47, 48] and hp-version adaptive finite element algorithms

version adaptive method, and the upper horizongtal x-axis represents the sparsely distributed final mesh used with the hp-version adaptive method.

Figure 8.5 shows the error distributions of the computed vibration modes in the mesh refinement and precision control stages of the hp-version adaptive FEM. The error distributions are shown for each mode, as determined with high precision with each method over the entire domain. To facilitate comparison, the error results of corresponding mode components are listed on the left and right sides, respectively. For a pre-specified error tolerance level $Tol = 10^{-6}$, the mode component ϕ^h in the h-version method exceeds the requirement. The refined meshes in the whole domain yield a solution error distribution of each mode component that is almost a running average of the tolerance Tol_h. After the order of the shape function of the element is increased by 1 (m was raised to 4), all the modes of the hp-version method strictly meet the error limitation requirements. Thus, the hp-version refinement may use fewer optimised meshes than h-version mesh refinement, and only one-step interpolation of the higher-order shape function can provide the eigensolutions satisfying the accuracy requirement.

8.6.2　Example 2: Thin-walled circular cylindrical shells

We here consider the free vibration of a thin-walled cylindrical shell with simply supported ends, using the same basic shell parameters as in Equation (8.32). The eigensolutions for different circumferential wave numbers $n = 1 \sim 15$ were obtained using our proposed method. Table 8.4 compares the computed frequency values with results of well-designed FEM[60] and dynamic stiffness method[61]. During the solution process, once the optimised meshes were obtained by h-version mesh refinement, the accuracy requirement was met after only a one-step interpolation of the higher-order shape function (m was raised to 4). By comparing the results for different circumferential wave numbers, the results computed by the proposed method agree well with the literature solutions, confirming its reliability for solving the frequencies of thin-walled circular cylindrical shells. These results are depicted graphically in Figure 8.6. The left and right axes represent, respectively, the frequency values and the frequency error values. With increasing circumferential wave number, the frequency clearly does not vary linearly; after an initial decrease, it then increases. The slight frequency error ($< 3 \times 10^{-4}$) results from the approximations used in the methods or models in the above cited studies.

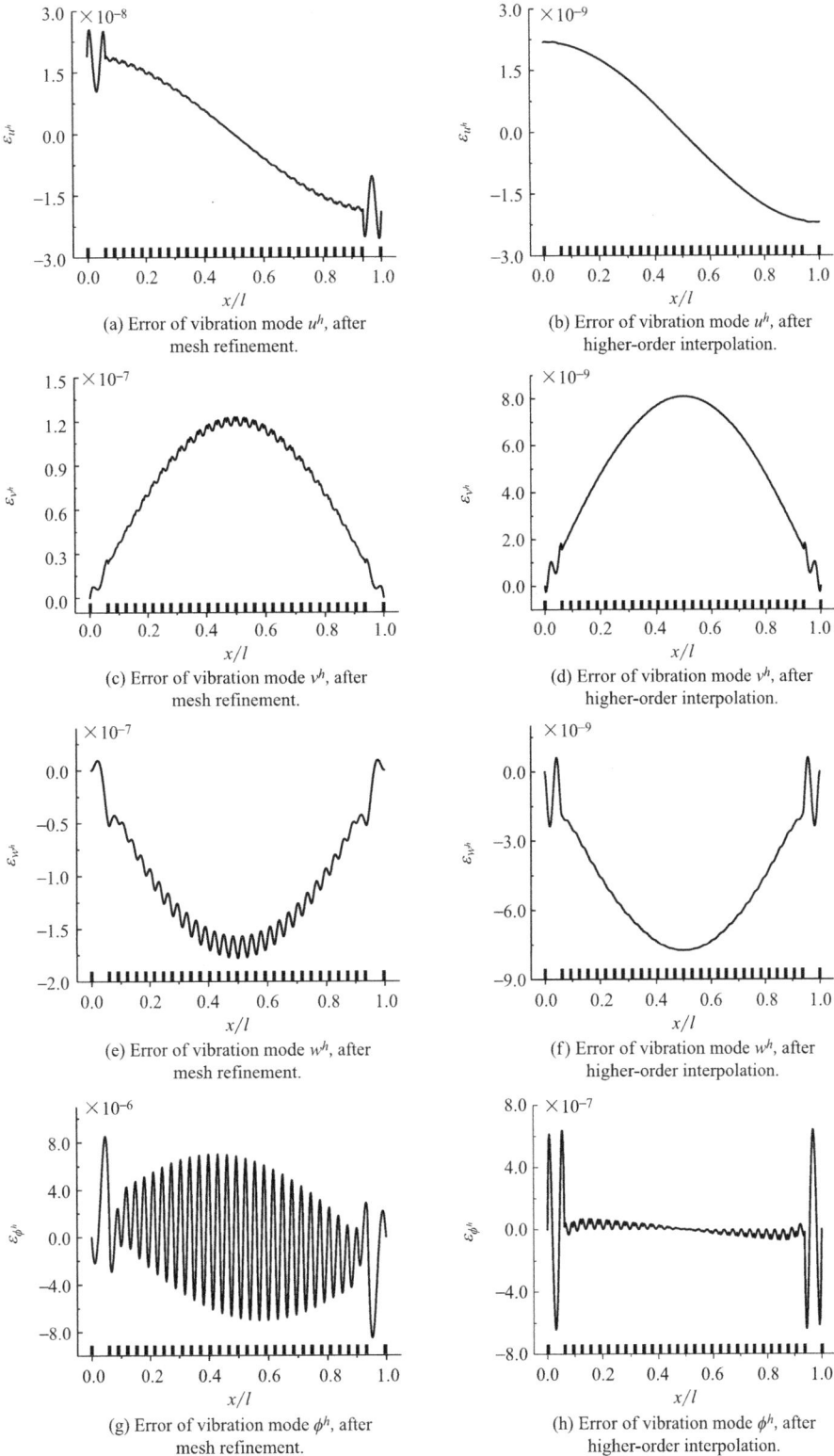

(a) Error of vibration mode u^h, after mesh refinement.

(b) Error of vibration mode u^h, after higher-order interpolation.

(c) Error of vibration mode v^h, after mesh refinement.

(d) Error of vibration mode v^h, after higher-order interpolation.

(e) Error of vibration mode w^h, after mesh refinement.

(f) Error of vibration mode w^h, after higher-order interpolation.

(g) Error of vibration mode ϕ^h, after mesh refinement.

(h) Error of vibration mode ϕ^h, after higher-order interpolation.

Figure 8.5   Example 1: error distributions of the computed vibration modes in mesh refinement and higher-order interpolation stages ($n=1$, *hp*-version adaptive FEM) (one)

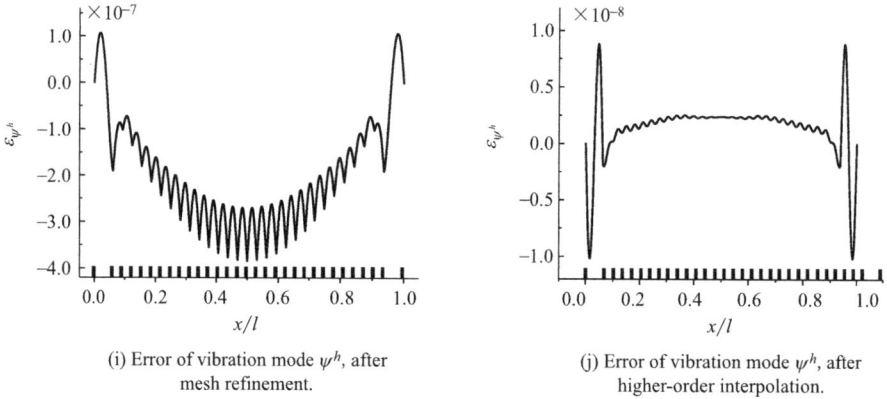

(i) Error of vibration mode ψ^h, after
mesh refinement.

(j) Error of vibration mode ψ^h, after
higher-order interpolation.

Figure 8.5　Example 1: error distributions of the computed vibration modes in mesh refinement
and higher-order interpolation stages ($n=1$, hp-version adaptive FEM) (two)

Computed results for the frequencies of Example 2: free vibration of thin-walled circular cylindrical shells

Table 8.4

Wave number	Frequencies				
n	ω^h	$\varepsilon_{\frac{a}{\omega}}$	$\varepsilon_{\frac{b}{\omega}}$	$\omega^{h\,①}$	$\omega^{h\,②}$
1	3270.95	1.07×10^{-4}	1.53×10^{-5}	3270.6	3270.9
2	1862.16	8.59×10^{-5}	3.22×10^{-5}	1862.0	1862.1
3	1101.86	5.44×10^{-5}	5.44×10^{-5}	1101.8	1101.8
4	705.710	2.69×10^{-5}	1.42×10^{-5}	705.9	705.7
5	497.484	3.21×10^{-5}	1.69×10^{-4}	497.5	497.4
6	400.081	4.74×10^{-5}	2.02×10^{-4}	400.1	400.0
7	380.702	5.24×10^{-6}	2.67×10^{-4}	380.7	380.6
8	416.700	0.00	2.39×10^{-4}	416.7	416.6
9	488.572	5.72×10^{-5}	1.47×10^{-4}	488.6	488.5
10	583.826	4.45×10^{-5}	4.45×10^{-5}	583.8	583.8
11	696.140	5.74×10^{-5}	5.74×10^{-5}	696.1	696.1
12	822.557	6.92×10^{-5}	6.92×10^{-5}	822.5	822.5
13	961.686	8.93×10^{-5}	8.93×10^{-5}	961.6	961.6
14	1112.85	1.35×10^{-4}	4.49×10^{-5}	1112.7	1112.8
15	1275.70	7.83×10^{-5}	7.83×10^{-5}	1275.8	1275.6

Sources: ①Results from paper [60], ②Results from paper [61].

Figure 8.7 shows the computed vibration modes and the corresponding final meshes
for the circumferential wave number $n=1$ using the hp-version adaptive FEM. The spatial
morphology of each mode component is shown over the entire cylindrical surface, together
with the distributions of the maxima and minima. The tick marks on the x-axis indicate
the distribution of the end nodes of each element. A denser mesh is used in the middle
region, where the vibration mode changes drastically. For this set of optimised meshes,

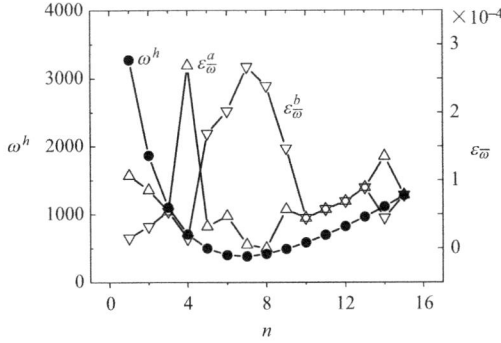

Figure 8. 6　Example 2: errors in the computed frequencies
(free vibration of thin-walled circular cylindrical shells)

the results meeting the accuracy requirements are obtained by a single increase in the order of the shape function. These results confirm the applicability of the proposed method for analysing thin-walled cylindrical shells.

8. 6. 3　Example 3: Circumferential wave number n

We here considerthe free vibration of moderately thick cylindrical shells simply supported at both ends with variable circumferential wave numbers n. The basic cylindrical shell parameters are as follows:

$$r=2 \text{ m}, \ h=0.2 \text{ m}, \ l=2 \text{ m}, \ J=h^3/12,$$

$$E=200 \text{ GPa}, \ \kappa=5/6, \ \mu=0.3, \ \rho=7800 \text{ kg/m}^3 \tag{8.33}$$

The proposed methodwas used to obtain the eigensolutions for different circumferential wave numbers, and the results of natural frequencies were transformed into dimensionless frequency values $\overline{\omega}^h = \omega^h l \sqrt{\rho(1+v)/E}$. In Table 8.5, these values are compared with other referenced methods, that is, a FE model based on the shape function along the circumferential direction of the shell[62], a three-dimensional linear elastic model[30], and a layered cylindrical shell model[28]. The discrepancies between our results and the other methods are small, confirming general consistency. The table data are shown graphically in Figure 8.8. The left and right axes represent, respectively, the frequency values and the frequency error values. As the circumferential wave number increases, the frequency does not vary linearly, but increases after an initial decrease. The frequency errors are less than 3×10^{-3}, and result from the approximations used in the methods or models in the cited studies.

Figure 8.9 shows the computed vibration modes and corresponding final meshes for circumferential wave numbers $n=1$ and 4, obtained using the *hp*-version adaptive FEM. The spatial morphology of each mode component is displayed over the entire cylindrical surface, together with the distributions of the maxima and minima. The tick marks on the x-axis indicate the distribution of the end nodes of each element. A denser mesh is used in

159

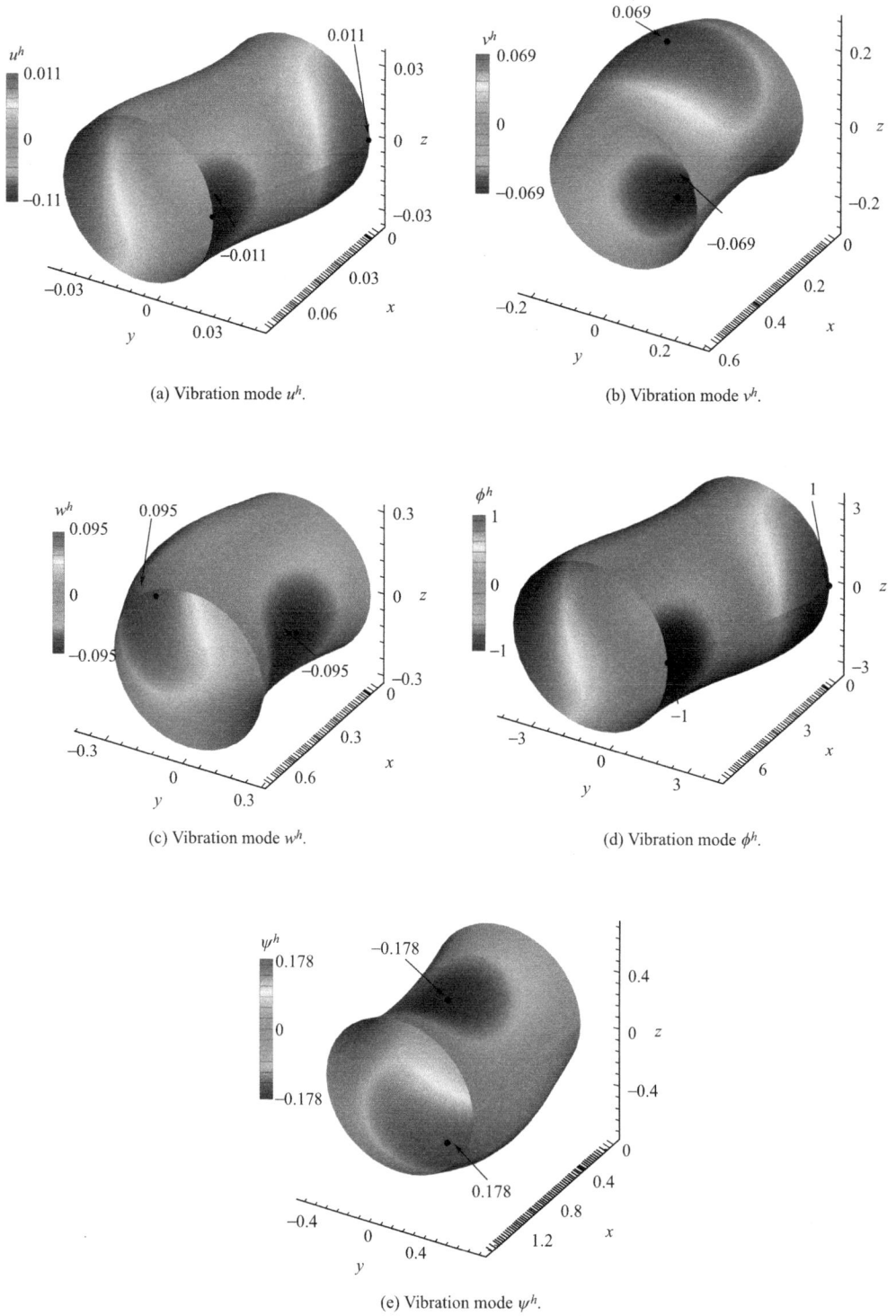

(a) Vibration mode u^h.

(b) Vibration mode v^h.

(c) Vibration mode w^h.

(d) Vibration mode ϕ^h.

(e) Vibration mode ψ^h.

Figure 8.7　Example 2: computed vibration modes and corresponding final meshes ($n=1$, of hp-version adaptive FEM)

Computed results for the dimensionless frequencies $\overline{\omega}^h = \omega^h l \sqrt{\rho\,(1+v)/E}$ of Example 3,

under different circumferential wave number n　　　　　Table 8.5

Wave number	Frequencies						
n	$\overline{\omega}^h$	$\varepsilon_{\overline{\omega}}^a$	$\varepsilon_{\overline{\omega}}^b$	$\varepsilon_{\overline{\omega}}^c$	$\overline{\omega}^{h}$ ①	$\overline{\omega}^{h}$ ②	$\overline{\omega}^{h}$ ③
1	1.063483	5.93×10^{-4}	5.35×10^{-4}	5.54×10^{-4}	1.06226	1.06238	1.06234
2	0.881473	4.55×10^{-4}	5.99×10^{-4}	5.61×10^{-4}	0.88233	0.88260	0.88253
3	0.806588	1.47×10^{-3}	1.68×10^{-3}	1.62×10^{-3}	0.80925	0.80963	0.80951
4	0.894386	2.31×10^{-4}	2.46×10^{-3}	2.39×10^{-3}	0.89877	0.89905	0.89893
5	1.115746	—	3.02×10^{-3}	2.99×10^{-4}	—	1.12216	1.12209

Sources: ①Results from paper [62], ②Results from paper [30], ③Results from paper [28].

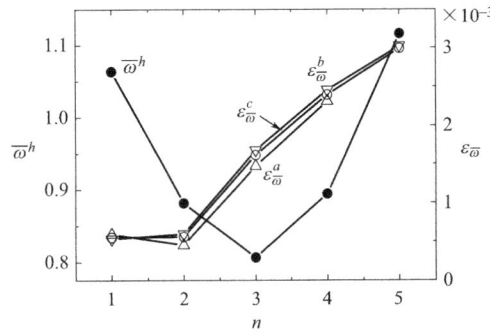

Figure 8.8　Example 3: computed frequencies for different circumferential wave numbers n

the middle region, where the vibration mode varies drastically. On this set of optimised meshes, the results meeting the accuracy requirements are obtained with a single increase in the order of the shape function. The adaptive mesh is the same for $n=1$ and $n=4$ because the circumferential change does not affect the radial vibration. The same final adaptive mesh was used here. These results indicate the suitability of the proposed method for analysing the shells with different circumferential wave numbers.

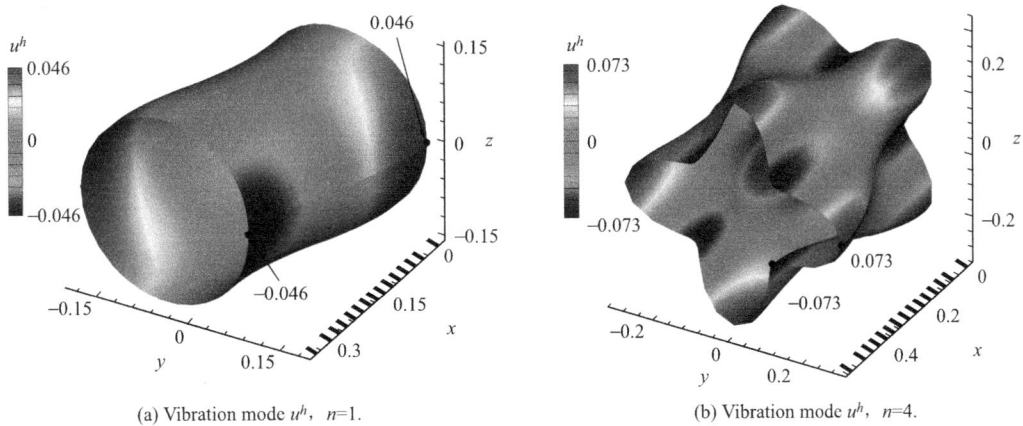

(a) Vibration mode u^h, $n=1$.　　　　　(b) Vibration mode u^h, $n=4$.

Figure 8.9　Example 3: computed vibration modes and corresponding final meshes for
different circumferential wave numbers n (one)

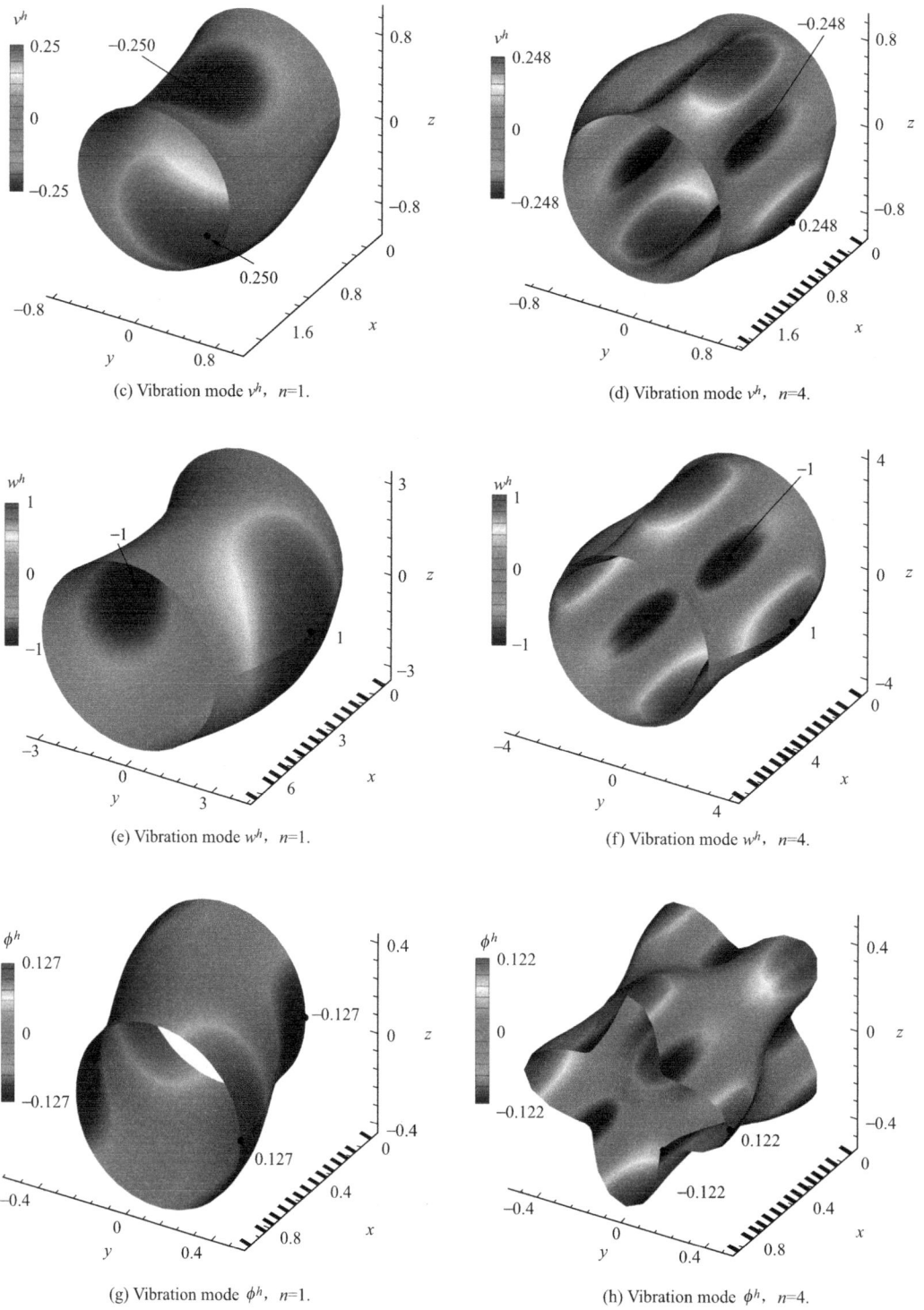

(c) Vibration mode v^h, $n=1$.

(d) Vibration mode v^h, $n=4$.

(e) Vibration mode w^h, $n=1$.

(f) Vibration mode w^h, $n=4$.

(g) Vibration mode ϕ^h, $n=1$.

(h) Vibration mode ϕ^h, $n=4$.

Figure 8.9　Example 3: computed vibration modes and corresponding final meshes for different circumferential wave numbers n (two)

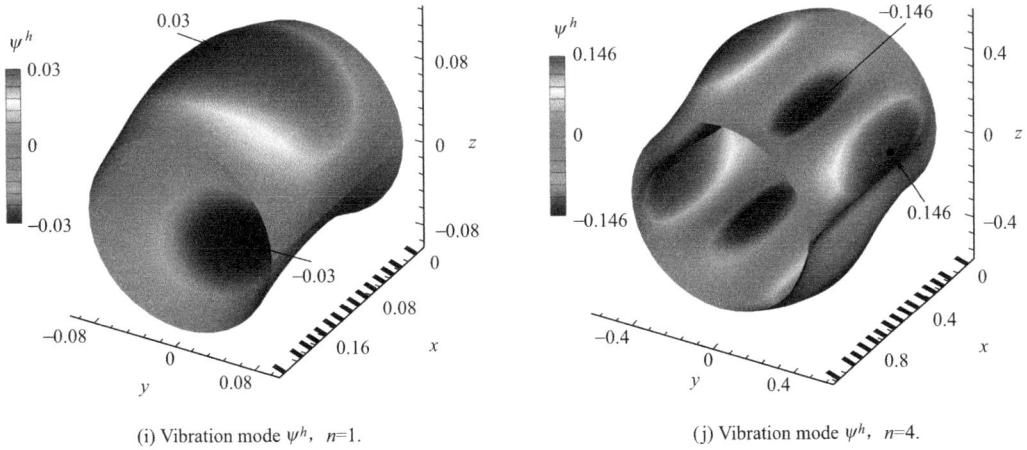

(i) Vibration mode ψ^h, $n=1$.　　　　　　(j) Vibration mode ψ^h, $n=4$.

Figure 8.9　Example 3: computed vibration modes and corresponding final meshes for different circumferential wave numbers n (three)

8.6.4　Example 4: Ratio of thickness to radius h/r

To verify the effectiveness of the proposed method for analysing thevariable geometrical thickness-to-radius ratio h/r, we consider the solutions of free vibration of a moderately thick cylindrical shell, simply supported at both ends, with $h/r=0.01$, 0.1, or 0.2. The circumferential wave number was set to $n=1$, and the other parameters were as follows:

$$h=4.8\text{m},\ l=24\text{m},\ J=h^3/12,$$
$$E=200\text{GPa},\ \kappa=5/6,\ \mu=0.3,\ \rho=7800\text{kg/m}^3 \tag{8.34}$$

The proposed methodwas used to solve for the eigensolutions corresponding to the different h/r values. Table 8.6 compares the resulting dimensionless natural frequencies $\overline{\omega}^h = (\omega^h h/\pi)\sqrt{\rho/G}$ with the results of other reference methods, namely an FE model based on the shape function along the circumferential direction of the shell[62] and a layered cylindrical shell model[28]. The discrepancies between these results are small, indicating a basic consistency between the methods. Figure 8.10 presents these data graphically. The left and right axes represent, respectively, the frequencies and the frequency errors. The frequency increases linearly with h/r and the frequency errors are less than 1×10^{-3}. These errors result from approximations in the methods or models used in the cited studies. In particular, when $h/r=0.1$, the relative error between the dimensionless value obtained by the present method and that obtained by Armenakas *et al.*[62] is greater than 0.001, whereas the computation results under other conditions are in good agreement.

Figure 8.11 shows the computed vibration modes and the corresponding final meshes corresponding to the two extreme conditions $h/r=0.01$ and 0.2 using the *hp*-version adaptive FEM. The figure shows the spatial morphology of each mode component over the entire cylindrical surface and displays the distributions of the maxima and minima. For

Computed results for the dimensionless frequencies $\overline{\omega}^h = (\omega^h h / \pi) \sqrt{\rho / G}$ of Example 4, for different thickness-to-radius ratios h/r　　　Table 8.6

Ratio of thickness to radius	Frequencies				
h/r	$\overline{\omega}^h$	$\varepsilon_{\overline{\omega}}^a$	$\varepsilon_{\overline{\omega}}^b$	$\overline{\omega}^{h\,①}$	$\overline{\omega}^{h\,②}$
0.01	0.057604	1.29×10^{-4}	1.29×10^{-4}	0.05774	0.05774
0.1	0.076418	1.15×10^{-3}	2.21×10^{-4}	0.07518	0.07618
0.2	0.107244	1.84×10^{-4}	1.84×10^{-4}	0.10704	0.10704

Sources: ①Results from paper [62], ②Results from paper [28].

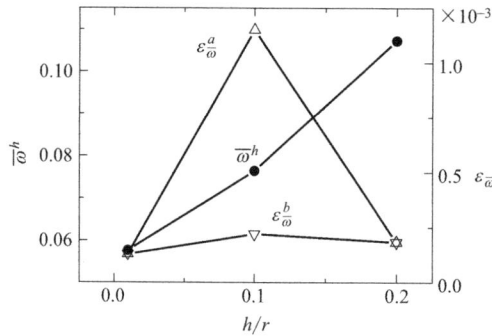

Figure 8.10　Example 4: computed frequencies for different thickness-to-radius ratios h/r

different h/r values, the vibration shape is exactly the same and consistent, the difference being the amplitude. The tick marks on the x-axis indicate the distribution of the end nodes of each element. A denser mesh is used in the middle region where the vibration mode varies significantly. On this set of optimised meshes, the results meeting the accuracy requirements are obtained by a single increase in the order of the shape function. The adaptive mesh is the same for $h/r = 0.01$ and 0.2 because the circumferential change does not affect the radial vibration, and the same final adaptive mesh is used here. These results indicate the applicability of the proposed method for analysing cylindrical shells with different h/r values.

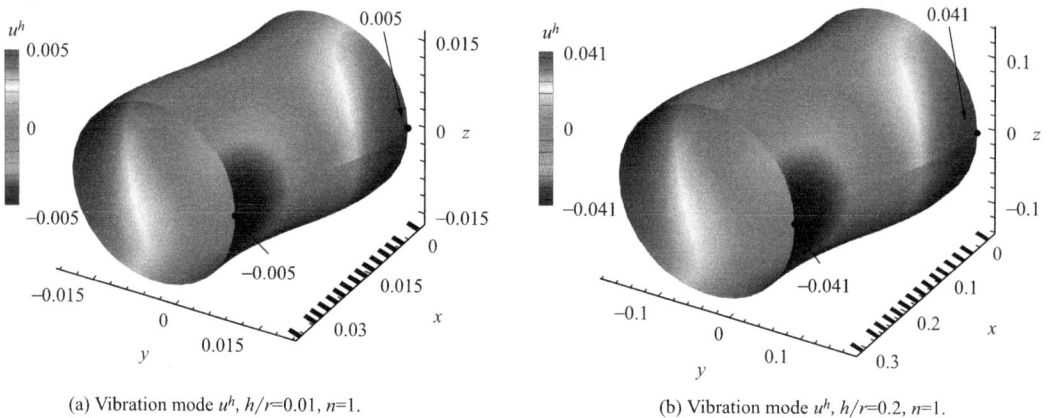

(a) Vibration mode u^h, $h/r=0.01$, $n=1$.　　　(b) Vibration mode u^h, $h/r=0.2$, $n=1$.

Figure 8.11　Example 4: computed vibration modes and corresponding final meshes for different thickness-to-radius ratios h/r (one)

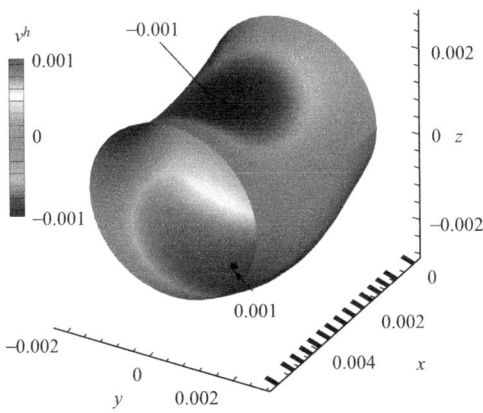

(c) Vibration mode v^h, h/r=0.01, n=1.

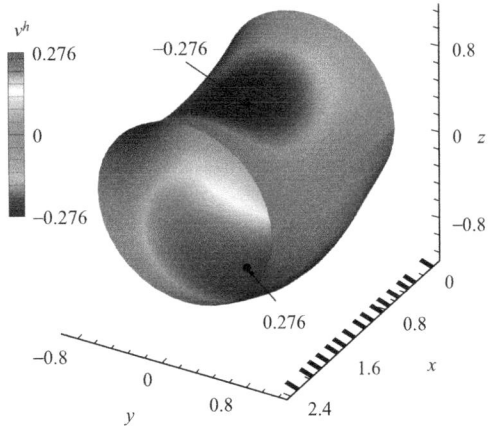

(d) Vibration mode v^h, h/r=0.2, n=1.

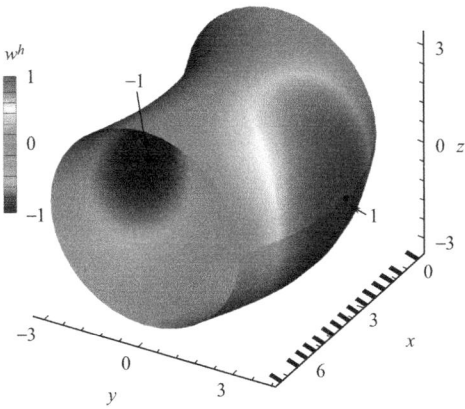

(e) Vibration mode w^h, h/r=0.01, n=1.

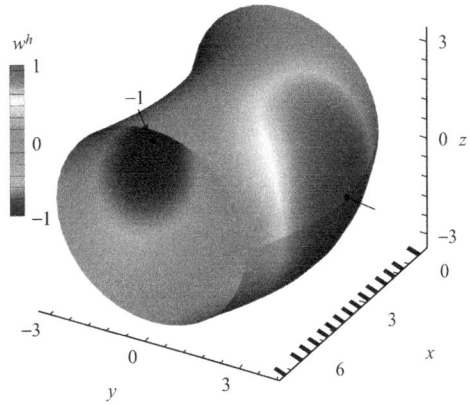

(f) Vibration mode w^h, h/r=0.2, n=1.

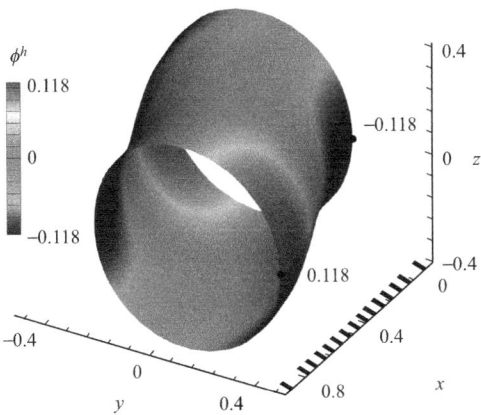

(g) Vibration mode ϕ^h, h/r=0.01, n=1.

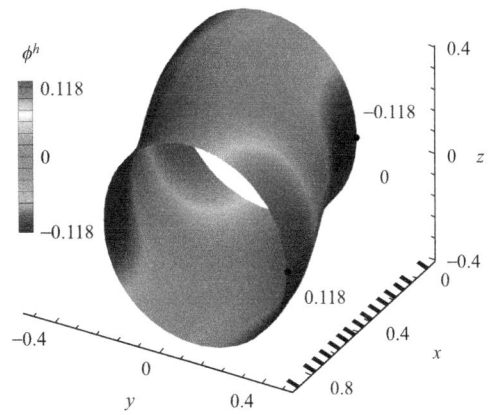

(h) Vibration mode ϕ^h, h/r=0.2, n=1.

Figure 8.11   Example 4: computed vibration modes and corresponding final meshes for different thickness-to-radius ratios h/r (two)

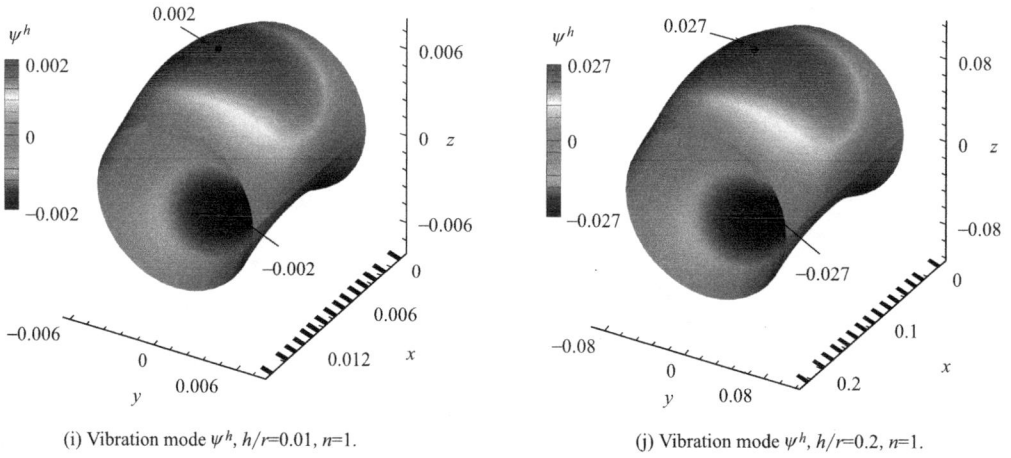

(i) Vibration mode ψ^h, h/r=0.01, n=1.

(j) Vibration mode ψ^h, h/r=0.2, n=1.

Figure 8.11　Example 4: computed vibration modes and corresponding final meshes for different thickness-to-radius ratios h/r (three)

8.6.5　Example 5: Ratio of thickness to length h/l

To verify the effectiveness of the proposed method for analysing thevariable geometrical thickness-to-length ratio h/l, we consider the solutions of free vibration of a moderately thick cylindrical shell, simply supported at both ends, with h/l=0.01, 0.1, 0.2, 0.4, 0.6, 0.8, or 1.0. The circumferential wave number of the cylindrical shell was set to n=0 (there is no change in the mode shape along the circumferential direction of the shell), and the other parameters were as follows:

$$r=2 \text{ m}, \ h=0.8 \text{ m}, \ J=h^3/12,$$
$$E=200 \text{ GPa}, \ \kappa=5/6, \ \mu=0.3, \ \rho=7800 \text{ kg/m}^3 \tag{8.35}$$

The proposed methodwas used to solve for the eigensolutions corresponding to the different h/l values. Table 8.7 compares the resulting dimensionless natural frequencies $\bar{\omega}^h=(\omega^h h/\pi)\sqrt{\rho/G}$ with the results of other above-mentioned reference methods, namely an FE model based on the shape function along the circumferential direction of the shell[62] and a layered cylindrical shell model[28]. The discrepancies between these results are small, which reflects the consistency between the methods. The results are displayed graphically in Figure 8.12. The left and right axes represent, respectively, the frequencies and the frequency errors. The frequency increases linearly with h/l and the frequency errors are less than 0.045. These errors result from the approximations in the methods or models in the cited studies. Especially when $h/l=0.1$, the relative error between the dimensionless value obtained by the present method and that obtained by Loy and Lam [28] is greater, while the computation results under other conditions are in good agreement.

166

Computed results for the dimensionless frequencies $\overline{\omega}^h = (\omega^h h / \pi)\sqrt{\rho/G}$ of

Example 5, under different thickness-to-length ratios h/l Table 8.7

Thickness-to-length ratio	Frequencies				
h/l	$\overline{\omega}^h$	$\varepsilon\frac{a}{\omega}$	$\varepsilon\frac{b}{\omega}$	$\overline{\omega}^{h\,①}$	$\overline{\omega}^{h\,②}$
0.01	0.016120	0.00	6.04×10^{-3}	0.01612	0.01002
0.1	0.152928	3.30×10^{-5}	4.81×10^{-2}	0.15289	0.10000
0.2	0.203862	9.03×10^{-4}	3.22×10^{-3}	0.20495	0.20000
0.4	0.278621	2.53×10^{-3}	2.49×10^{-3}	0.27540	0.27544
0.6	0.422104	1.33×10^{-3}	1.24×10^{-3}	0.42022	0.42035
0.8	0.597491	1.62×10^{-3}	1.77×10^{-3}	0.60009	0.60033
1.0	0.784201	4.76×10^{-3}	4.99×10^{-3}	0.79274	0.79314

Source: ①Results from paper [62], ②Results from paper [28].

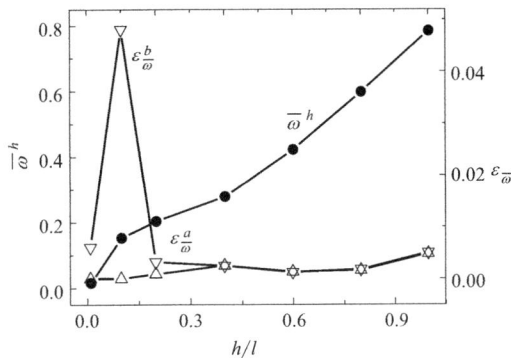

Figure 8.12   Example 5: computed frequencies
for different thickness-to-length ratios h/l

Figure 8.13 shows the computed vibration modes and corresponding final meshes of the two extreme conditions of $h/l=0.01$ and 1 using *hp*-version adaptive FEM. The figure shows the spatial morphology of each mode component over the entire cylindrical surface and displays the distributions of the maxima and minima. The components of the vibration mode with a value of 0 are not shown in the figure. For different values of h/l, the vibration shape is exactly the same and consistent, the difference being the amplitude. When $h/l=0.01$, the maximum amplitude 1 of the mode appears in mode component u^h, whereas when $h/l=1$, it appears in mode component w^h, which indicate the influence of shell geometry on vibration modes. The tick marks on the x-axis indicate the distribution of the end nodes of each element. A denser mesh is used in the middle region, where the vibration mode varies significantly. There are 16 elements for $h/l=0.01$, and 14 elements for $h/l=1$, that are reasonably optimised according to the changing vibration modes. On this set of optimised meshes, the results meeting the accuracy requirements were obtained by a single increase in the order of the shape function. These results indicate the applicability of the proposed method for analysing cylindrical shells with different h/l values.

167

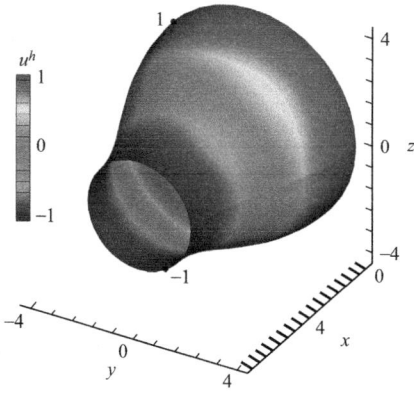

(a) Vibration mode u^h, h/l=0.01, n=0

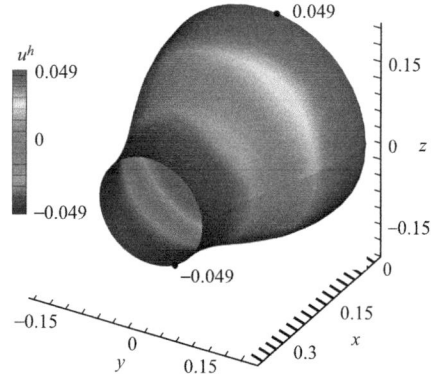

(b) Vibration mode u^h, h/l=1, n=0

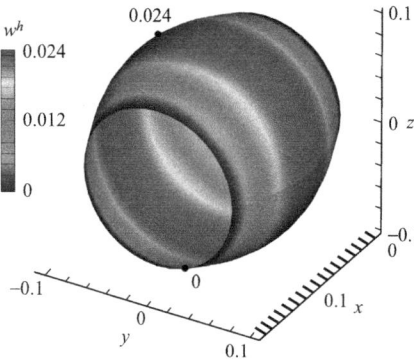

(c) Vibration mode w^h, h/l=0.01, n=0

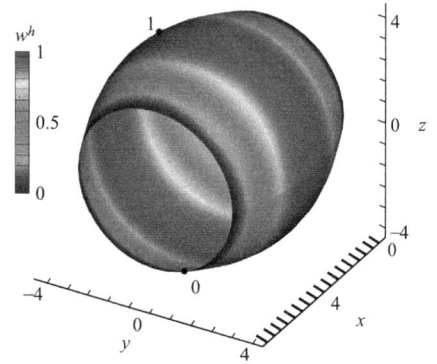

(d) Vibration mode w^h, h/l=1, n=0

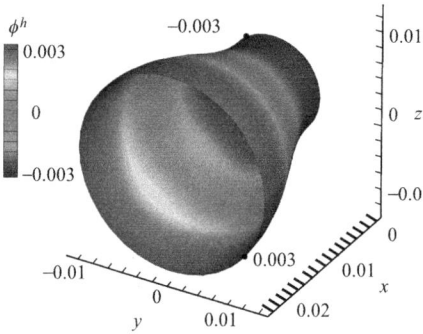

(e) Vibration mode ϕ^h, h/l=0.01, n=0

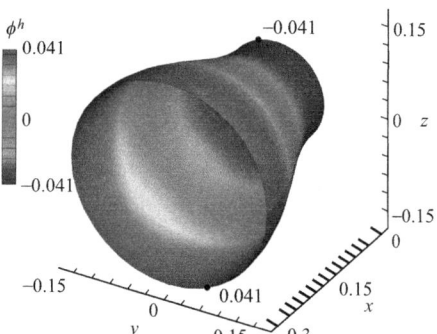

(f) Vibration mode ϕ^h, h/l=1, n=0

Figure 8.13　Example 5: computed vibration modes and corresponding final meshes for different thickness-to-length ratios h/l

8.7　Conclusions

This study proposed a novel hp-version adaptive FEM to accurately and efficiently solve for the free vibration solutions of moderately thick cylindrical shells. The algorithm comprehensively combines the *h*- and *p*-version adaptive FE techniques to obtain an optimised adaptive mesh and a high-precision solution satisfying a pre-specified error tolerance condition. Challenging issues regarding variable geometrical factors, such as the thickness, circumferential wave number, radius, and length, were overcome and solved. Representative numerical examples verified the efficiency and reliability of the proposed method. The conclusions are summarised as follows:

(1) An *hp*-version adaptive FE algorithm involving error homogenisation and higher-order interpolation was proposed. It combined the *h*-version mesh refinement and *p*-version higher-order shape functions for interpolation. The *hp*-version refinement uses fewer optimised meshes than *h*-version mesh refinement, and only one-step interpolation of the higher-order shape function can provide the eigensolutions satisfying the accuracy requirement.

(2) The *hp*-version adaptive FE analysis for models of circular cylindrical shells with variable wall thickness was consistently implemented. For moderately thick and thin-walled shells, the computed solutions of the frequencies and mode shapes obtained agree well with results from other high-precision methods (such as the dynamic stiffness method, serving as an analytical technique). The applicability of this method is confirmed.

(3) The non-uniform mesh generation and higher-order interpolation of shape functions used in the proposed algorithm are suitable for addressing the problem of complex frequencies and modes caused by changes in structural geometry. The optimal non-uniform distribution mesh can be obtained for solving the solutions of continuous order frequency and mode shapes of moderately thick cylindrical shells with different circumferential wave numbers, thickness-to-radius ratios h/r, and thickness-to-length ratios h/l. The results show that the adaptive method automatically generates an optimised FE mesh in the region where the modal (vibration mode) solution changes sharply, while the other regions are relatively sparse. The proposed method effectively yields high-precision solutions that meet the pre-specified error tolerance condition.

This algorithm can be extended toaddress general eigenproblems and geometric forms of structures to find the solutions for the frequency and mode quickly and efficiently. In particular, for general high-dimensional structural problems (e. g. , two-dimensional plates and shells and three-dimensional solid structure eigenvalue problems with geometric changes) and vibration disturbance-induced by local damage, the solutions are strongly mesh-dependent, and these problems require efficient solutions. This aspect will need further development of the proposed *hp*-version adaptive FE algorithm.

References

[1] Dey T, Ramachra. Non-linear vibration analysis of laminated composite circular cylindrical shells [J].

Composite Structures, 2017, 163: 89-100.

[2] Donnell L H. Stability of thin-walled tubes under torsion [D]. Pasadena: California Inst of Tech, 1933.

[3] Sanders J, Lyell J. An improved first-approximation theory for thin shells [J]. NASA Report NASA-TR-R24, 1959.

[4] Flügge W. Stresses in Shells: [M]. Springer Berlin Heidelberg, 1973.

[5] Love A E H. A treatise on the mathematical theory of elasticity [M]. Cambridge: Cambridge University Press, 2013.

[6] Confalonieri F, Ghisi A, Perego U, et al. 8-Node solid-shell elements selective mass scaling for explicit dynamic analysis of layered thin-walled structures [J]. Computational Mechanics, 2015, 56 (4): 585-559.

[7] Messina A, Soldatos K P. Ritz-type dynamic analysis of cross-ply laminated circular cylinders subjected to different boundary conditions [J]. Journal of Sound and Vibration, 1999, 227 (4): 749-768.

[8] Sivadas K R, Ganesan N. Free vibration and material damping analysis of moderately thick circular cylindrical shells [J]. Journal of Sound and Vibration, 1994, 172 (1): 47-61.

[9] Ahad F, Shi D, Hian Z, et al. Free vibration analysis of moderately thick isotropic homogeneous open cylindrical shells using improved Fourier series method [J]. Journal of Vibroengineering, 2017, 19 (5): 3679-3693.

[10] Shi D, He D, Wang Q, et al. Free vibration analysis of closed moderately thick cross-ply composite laminated cylindrical shell with arbitrary boundary conditions [J]. Materials (Basel), 2020, 13 (4).

[11] Thinh T I, Nguyen M C. Dynamic stiffness matrix of continuous element for vibration of thick cross-ply laminated composite cylindrical shells [J]. Composite Structures, 2013, 98: 93-102.

[12] Lam K Y, Loy C T. Analysis of rotating laminated cylindrical shells by different thin shell theories [J]. Journal of Sound and Vibration, 1995, 186 (1): 23-35.

[13] Li X. A new approach for free vibration analysis of thin circular cylindrical shell [J]. Journal of Sound and Vibration, 2006, 296 (1-2): 91-98.

[14] Pan Z, Li X, Ma J, et al. A study on free vibration of a ring-stiffened thin circular cylindrical shell with arbitrary boundary conditions [J]. Journal of Sound and Vibration, 2008, 314 (1/2): 330-342.

[15] Zhang X M Liu G R, Lam K Y. Frequency analysis of cylindrical panels using a wave propagation approach [J]. Applied Acoustics, 2001, 62 (5): 527-543.

[16] Zhang X M. Parametric analysis of frequency of rotating laminated composite cylindrical shells with the wave propagation approach [J]. Computer Methods in Applied Mechanics and Engineering, 2002, 191 (19-20): 2057-2071.

[17] Li X. Study on free vibration analysis of circular cylindrical shells using wave propagation [J]. Journal of Sound and Vibration, 2008, 311 (3-5): 667-682.

[18] Nevenka K, Marija N D. Dynamic stiffness-based free vibration study of open circular cylindrical shells [J]. Journal of Sound and Vibration, 2020, 486 (10): 1884-2021.

[19] Tong Z, Ni Y, Zhou Z, et al. Exact solutions for free vibration of cylindrical shells by a symplectic approach [J]. Journal of Vibration Engineering and Technologies, 2018, 6 (2): 107-115.

[20] Gao R, Sun X, Liao H, et al. Symplectic wave-based method for free and steady state forced vibration analysis of thin orthotropic circular cylindrical shells with arbitrary boundary conditions [J]. Journal of Sound and Vibration, 2021, 491.

[21] Chen X, Ye K. Comparison study on the exact dynamic stiffness method for free vibration of thin and moderately thick circular cylindrical shells [J]. Shock and Vibration, 2016, 2016 (1/14).

170

[22] Setoodeh A R, and Karami G. Static, free vibration and buckling analysis of anisotropic thick laminated composite plates on distributed and point elastic supports using a 3-D layer-wise FEM [J]. Engineering Structures, 2004, 26 (2): 211-220.

[23] Yan Y, Pagani A, , et al. Exact solutions for free vibration analysis of laminated, box and sandwich beams by refined layer-wise theory [J]. Composite Structures, 2017, 175 (28-45).

[24] Ferreira A J M, Carrera E, Cinefra M, et al. Radial basis functions collocation for the bending and free vibration analysis of laminated plates using the Reissner-Mixed Variational Theorem [J]. European Journal of Mechanics-A/Solids, 2013, 39 (104-112).

[25] Pramod A L N, Natarajan S, Ferreira A J M, et al. Static and free vibration analysis of cross-ply laminated plates using the Reissner-mixed variational theorem and the cell based smoothed finite element method [J]. European Journal of Mechanics-A/Solids, 2017, 62 (14-21).

[26] Karczub D G. Expressions for direct evaluation of wave number in cylindrical shell vibration studies using the Flügge equations of motion [J]. Journal of the Acoustical Society of America, 2006, 119 (6).

[27] Singal R K, Williams K. A theoretical and experimental study of vibrations of thick circular cylindrical shells and rings [J]. Journal of Vibration and Acoustics, 1988, 110 (4): 533-537.

[28] Loy C T, Lam K Y. Vibration of thick cylindrical shells on the basis of three-dimensional theory of elasticity [J]. Journal of sound and Vibration, 1999, 226 (4): 719-737.

[29] Khalili S M R, Davar A, Malekzadeh F K, et al. Free vibration analysis of homogeneous isotropic circular cylindrical shells based on a new three-dimensional refined higher-order theory [J]. International Journal of Mechanical Sciences, 2012, 56 (1): 44951.

[30] Soldatos K P, Hadjigeorgiou V P. Three-dimensional solution of the free vibration problem of homogeneous isotropic cylindrical shells and panels [J]. Journal of Sound & Vibration, 1990, 137 (3): 369-384.

[31] Viswanathan K K, Kim K S, Lee J H, et al. Free vibration of multi-layered circular cylindrical shell with cross-ply walls, including shear deformation by using spline function method [J]. Journal of Mechanical Science and Technology, 2009, 22 (11): 2062-2075.

[32] Rawat A, Matsagar V A, Nagpal A K, et al. Free vibration analysis of thin circular cylindrical shell with closure using finite element method [J]. International Journal of Steel Structures, 2020, 20 (1): 175-193.

[33] Gerasimov T, Stein E, Wriggers P, et al. Constant-free explicit error estimator with sharp upper error bound property for adaptive FE analysis in elasticity and fracture [J]. International Journal for Numerical Methods in Engineering, 2015, 101 (2): 79-126.

[34] Yuan S, Wang Y, Ye K, et al. An adaptive FEM for buckling analysis of nonuniform Bernoulli-Euler members via the element energy projection technique [J]. Mathematical Problems in Engineering, 2013, 2013 (1/6).

[35] Zienkiewicz O C. The background of error estimation and adaptivity in finite element computations [J]. Computer Methods in Applied Mechanics and Engineering, 2006, 195 (4/6): 207-213.

[36] Bespalov A, Haberl A, Praetorius D, et al. Adaptive FEM with coarse initial mesh guarantees optimal convergence rates for compactly perturbed elliptic problems [J]. Computer Methods in Applied Mechanics and Engineering, 2017, 317 (15): 318-340.

[37] Zienkiewicz O C, Zhu J Z. A simple error estimator and adaptive procedure for practical engineering analysis [J]. International Journal for Numerical Methods in Engineering, 1987, 24 (2): 337-357.

[38] Wang Y, Ju Y, Chen J, et al. Adaptive finite element-discrete element analysis for the multistage supercritical CO2 fracturing of horizontal wells in tight reservoirs considering pre-existing fractures and thermal-hydro-mechanical coupling [J]. Journal of Natural Gas Science and Engineering, 2019, 195 (4/6): 207-213.

[39] Douglas J, Dupont T. Galerkin approximations for the two point boundary problems using continuous piecewise polynomial spaces [J]. Numerical Mathematics, 1974, 22 (2): 99-109.

[40] Arthurs C J, Bishop M J, Kay D, et al. Efficient simulation of cardiac electrical propagation using high-order finite elements II: Adaptive p-version [J]. Journal of Computational Physics, 2013, 253: 443-470.

[41] Bardell N S, DunsdonJ, M, et al. On the free vibration of completely free open cylindrically curved isotropic shell panels [J]. Journal of Sound and Vibration, 1997, 207 (5): 647-669.

[42] Tews R, Rachowicz W. Application of an automatic hp adaptive finite element method for thin-walled structures [J]. Computer Methods in Applied Mechanics and Engineering, 2009, 198 (21-26): 1967-1984.

[43] Tóth B. Hybridized dual-mixed hp-finite element model for shells of revolution [J]. Computers and Structures, 2019, 218: 123-151.

[44] Arndt M, Machado R D, Scremin A, et al. An adaptive generalized finite element method applied to free vibration analysis of straight bars and trusses [J]. Journal of Sound & Vibration, 2010, 329 (6): 659-672.

[45] Arndt M, Machado R D, Scremin A, et al. Accurate assessment of natural frequencies for uniform and non-uniform Euler-Bernoulli beams and frames by adaptive generalized finite element method [J]. Engineering Computations, 2016, 33 (5): 1586-1609.

[46] Stein E, Seifert B, Ohnimus S, et al. Adaptive finite element analysis of geometrically non-linear plates and shells, especially buckling", International Journal for Numerical Methods in Engineering, Vol. 37 No. 15, pp. 2631-2655. [J]., 2010.

[47] Wang Y, Ju Y, Zhuang Z, et al. Adaptive finite element analysis for damage detection of non-uniform Euler-Bernoulli beams with multiple cracks based on natural frequencies [J]. Engineering Computations, 2018, 35 (3): 1203-1229.

[48] Wang Y. An h-version adaptive FEM for eigenproblems in system of second order ODEs: Vector Sturm-Liouville problems and free vibration of curved beam [J]. Engineering Computations, 2020, 37 (1): 1210-1225.

[49] Zienkiewicz O C, Taylor R L, Zhu, et al. The finite element method: its basis and fundamentals (7th Edition) [M]. Oxford UK: Elsevier (Singapore), 2015.

[50] Wilkinson J H. The Algebraic Eigenvalue Problem [J]. Clarendon Press Oxford. , 1965.

[51] Wilkinson J H, Reinsch C. Linear algebra, handbook for automatic computation [J]. Springer-Verlag New York. , 1971,

[52] Bérard P, Helffer B. Sturm's theorem on zeros of linear combinations of eigenfunctions [J]. Expositiones Mathematicae, 2020, 38 (1): 27-50.

[53] Thomas J M. Sturm's theorem for multiple roots [J]. National Mathematics Magazine, 1941, 15 (8): 391-394.

[54] Ioakimidis N I. Deciding in elasticity problems by using Sturm's theorem [J]. Computers and structures, 1996, 58 (1): 123-131.

[55] Wiberg N E, Bausys R, Hager P, et al. Adaptive h-version eigenfrequency analysis [J]. Computers

and Structures, 1999, 71 (5): 565-584.

[56] Wiberg N E, Bausys R, Hager P, et al. Improved eigenfrequencies and eigenmodes in free vibration analysis [J]. Computers & Structures, 1999, 73 (1/5): 79-89.

[57] Schillinger D, Rank E. An unfitted *hp*-adaptive finite element method based on hierarchical B-splines for interface problems of complex geometry [J]. Computer Methods in Applied Mechanics and Engineering, 2011, 200 (4748): 3358-3380.

[58] Bao G, Hu G, Liu D, et al. An *h*-adaptive finite element solver for the calculations of the electronic structures [J]. Journal of Computational Physics, 2012, 231 (14): 4967-4979.

[59] Clough R W, Penzien J. Dynamics of Structures: Second edition [J]. New York: McGraw-Hill, 1993.

[60] Leissa A W. Vibration of shells [Z]. Scientific and Technical Information Office National Aeronautics and Space Administration. , 1973.

[61] Chen X, Ye K. Analysis of free vibration of moderately thick circular cylindrical shells using the dynamic stiffness method [J]. Engineering Mechanics, 2016, 33 (9): 40-48.

[62] Armenakas A E, Gazis D S, Herrmann G, et al. Free vibrations of circular cylindrical shells [J]. Pergamon Press: Oxford, 1969.

Chapter 9
Adaptive finite element method for vibration disturbance of cracked cylindrical shells

9. 1　Introduction

　　The dynamic analysis of structures and elastomers is an important basis for the study of structural earthquake resistance and rock-induced earthquakes[1, 2]. As a supporting structure or storage cavity, the cylindrical shell is widely used in structural engineering, rock engineering, and aerospace engineering, and studying the dynamic characteristics of the structure, such as vibration, instability, and buckling, is of great significance for studying and judging its failure behaviour [3]. Nondestructive experimental methods for computing the buckling load of imperfection-sensitive thin-walled structures are one of the most important techniques for the validation of new structures and numerical models of large-scale aerospace structures [4]. In the study of structural dynamics, natural frequency and vibration mode are used as mechanical response parameters to analyse dynamic characteristics [5, 6], and have become key research areas. These frequency and vibration modes are also regarded as eigenvalues and eigenfunctions of mathematical eigenvalue problems, which are solved. The circular cylindrical shell has, among other characteristics, clear force, symmetrical structure, and relatively simple manufacturing process [7]. Therefore, the free vibration of a cylindrical shell has been of interest to many researchers [8]. Consequently, an accurate analysis of the free vibration of circular cylindrical shells is highly valued in practice and research. At present, the traditional thin shell theory is often used to study shell problems. Based on the Kirchhoff-Love assumption [9] and ignoring the transverse shear deformation, the theory introduces some errors to a shell structure with small shear stiffness (that is, prone to significant transverse shear deformation). In addition, it underestimates deflections and overestimates the frequencies [10]. As many applications of toroidal shells are moderately thick or thick, it is imperative and desirable to establish effective theories appropriate for their analysis [11]. In addition, the increased wall thickness of plate and shell structures in practical engineering is often beyond the application range of the thin-walled theory, and the influence of transverse shear deformation should also be considered. This study introduced a moderately thick circular cylindrical shell. Compared with the thin shell free vibration theory, the moderately thick shell free vibration theory considers the influence of transverse shear deformation and moment of inertia, which makes the solution more reliable.

　　Initial defects and factors such as high strength work and long-term use can damage the circular cylindrical shell structure, and cracks are one .of the most common defects in most structures. Regarding the power characteristics of the structure, the presence of cracks reduces the original frequency and pushes the resonance band, causing structural vibration and stress strain aggravation, which increases the length of cracks. Cracks in a structural element in the form of initial defects or local imperfection within the material, or caused by fatigue, buckling or stress concentration, certainly impact the structural integrity[12, 13]. This creates a vicious circle that affects the reliability of the structure and directly threatens its safety. It is necessary to study the dynamic characteristics of a cracked structure, particularly the characteristics of free vibration. Therefore, in structural design and engineering applications it is important to study the mechanical properties of cracks and to clarify the ultimate bearing capacity of structures containing cracks. Cracks are one of the most common defects in engineering structures; their existence inevitably leads to a decrease in the ultimate bearing capacity of structures or components, and structural fractures will lead to significant economic losses [14].

In the study of the vibration of cracked cylindrical shells, it is necessary to analyse the influence ofdifferent crack positions, sizes, and numbers. Compared with functionally graded beams with a single crack, beams with two or more cracks have lower frequency values [15]. To investigate the influences of the crack depth and position of each crack on the vibration mode and natural frequencies of a simply supported beam, Yoon et al. [16] derived the equation of motion using Hamilton's principle and analysed using a numerical method. To learn about the effects of the position and depth of each crack on the natural frequency of a simply supported double-cracked beam, Jafari et al. [17] also conducted a free vibration analysis of cylindrical shells with circumferential stiffeners. To understand the free vibrations of circular cylindrical shells of piecewise constant thickness when circular cracks of constant length are controlled, Jaan et al. [18] presented an approximate method for the vibration analysis of stepped shells accounting for the influence of cracks located at the re-entrant corners of steps. To analyse the coupled vibration feature of a fluid-filled cylindrical shell with a circumferential surface crack, Jin et al. [19] also analysed the effects of crack depth, crack location, and boundary conditions on the cylindrical shell's modal power flow index. To determine the effects of various geometric properties on the vibration and damping factors of circular cylindrical shells, Sivadas et al. [6] used an improved shell theory with shear deformation and rotatory inertia. To implement a straightforward computer simulation under different shell theories and boundary conditions, Lee et al. [20] used the Rayleigh-Ritz method to analyse the free vibration of a circular cylindrical shell on a dynamic model. To obtain solutions for the problem of free non-axisymmetric vibration of stepped circular cylindrical shells with cracks, Roots [21] prescribed the influence of circular cracks with constant depth on the vibration of the shell with the aid of a matrix of local flexibility. Most research in the area of vibration analysis and crack detection of cracked structures has focused on beams [22]. To determine the effect of the development and propagation of cracks on cracked cylindrical shells with various parameters, Dehghani et al. [23] examined the effect of a complete penetration non-propagating macro-crack damage on natural vibration frequencies and mode shapes. However, there is a lack of research on the influence of vibration on the location, depth, number, and distribution of damage with annular cracks in cylindrical shells. If these problems are not clear, it is difficult to evaluate the vibration disturbance behaviour of cylindrical shells according to crack damage, and it is more difficult to further identify crack damage information according to vibration.

Some theoretical methods, experiments, and numerical computations have been developed to study the free vibration of medium-thick cylindrical shells with cracks. To identify the crack location and depth, Zhu et al.[24] presented an approach to analyse the wave and vibrational power flow characteristics in cracked cylindrical shell structures. To conduct a modal analysis of the vibration response of a cracked fluid-filled cylindrical shell, Zheng[25] derived a high-order partial differential equation of thin shell motion. In the development of a research on the free vibration of medium-thick cylindrical shells, the dynamic stiffness method was introduced in the study of moderately thick circular cylindrical shells[26, 27]. The free vibration of cylindrical shells under homogeneous boundary conditions was studied based on the theory of Flugge[28] and elastic thin shells[29]. To study the effect of cracks on the natural frequencies and mode shapes of cracked beams, Li[30] proposed an exact approach for free vibration analysis of a non-uniform beam with an arbitrary number of cracks and concentrated masses. To study the influence of a change in shell thickness on the distribution of its natural frequencies, Grigorenko et al.[31] introduced the spline collocation method for finding the frequencies of free vibrations of circular closed cylindrical shells of variable thickness in the circumferential direction. Grigorenko et al.[32] used a holographic interferometry technique to determine the frequencies of the free vibrations of isotropic circular cylindrical shells. To improve the computational efficiency and accuracy, Pellicano[33] proposed a general framework to analyse the vibration of circular cylindrical shells, both in the case of linear and nonlinear vibrations. At present, there is a lack of a reliable and high-precision analysis of the frequency and mode of vibration of the disturbed mode of cylindrical shells with crack damage.

For the numerical models and analysis of eigensolutions, the finite element method is widely used to solve the free vibration of moderately thick circular cylindrical shells with complex structures and boundary conditions. The adaptive algorithm for the finite element method has become an important method for optimising the mesh and improving the solution accuracy[34-39]. It mainly includes the p-adaptive method[40] for improving the element order, h-adaptive method for increasing the mesh density[41, 42], and hp-adaptive method[43-45], which combines the above two methods. Through high-performance computations, an adaptive analysis of the finite element method makes it possible to reliably solve challenging issues such as eigenvalue problems, damage, and fracture[46-48]. An error analysis of local imperfections based on numerical solutions[49, 34] was proposed, and the energy norm measurement method was used to compute the elements with the largest local imperfections. Oden and Ainsworth[50, 51] also made significant contributions to the analysis of residual errors. In the study of the free vibration of cylindrical shells, the adaptive finite element method can effectively provide higher precision solutions for the analysis. To predict crack-induced natural frequency changes, an accurate and efficient method for analysing the vibration characteristics of cylindrical shells with a part-through crack was proposed[52]. The non-uniform refined mesh in a high-performance adaptive analysis process is a crucial factor for efficient computation, in which the minimum number of elements and optimised distribution of nodes are used to reduce the solution errors quickly and derive high-precision solutions[53-55]. The displacement superconvergent patch recovery method of finite elements[56, 57, 47, 42] was proposed for establishing the mesh refinement procedure for computing high-precision modal shapes. The adaptive finite element method was used to analyse the free vibration of beams with multiple cracks and damage detection[58]. This study extends this method to the vibration disturbance problem of moderately thick circular cylindrical shells with crack damage.

The remainder of this chapter is organised as follows. In Section 9. 2, we present the differential equations describing the free vibration of moderately thick circular cylindrical shell. The damage characterisation method for circumferential cracks in a circular cylindrical shell is introduced in Section 9. 3. In Sections 9. 4 and 9. 5, the key techniques used in the adaptive analysis procedure, such as error estimation for eigenfunctions and h-version mesh refinement, are presented. Representative numerical examples are presented in Section 9. 6 to demonstrate the performance of the proposed method and algorithm. Finally, the main conclusions are summarised in Section 9. 7.

9. 2   Differential equations describing the free vibration of moderately thick circular cylindrical shell

The geometrical model, coordinate systems, and parameters describing the linear elastic moderately thick circular cylindrical shell with circumferential cracks investigated in this study are shown in Figure 9. 1. The rotation axis is denoted by ox. The local coordinate system at point A on the mid-plane is defined by $\bar{\alpha}\bar{\beta}\bar{\gamma}$, where $\bar{\alpha}$ points along the tangential direction of the meridian (the direction of the shell axis), $\bar{\beta}$ points along the tangential direction of the weft circle, and $\bar{\gamma}$ points along the normal direction. There is a

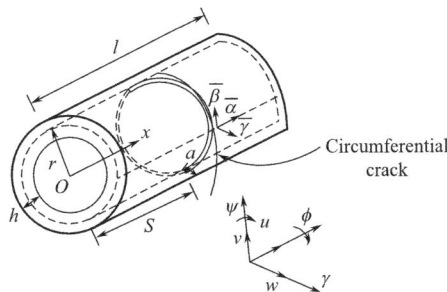

Figure 9. 1   Geometrical model, coordinate systems, and symbols describing a moderately thick circular cylindrical shell with circumferential crack

179

circumferential surface micro-crack in the cylindrical shell, which is located at length s along the rotation axis, and the depth of the crack is along the thickness of the shell. The five independent displacements of the shell are the linear displacements u, v, and w along the $\bar{\alpha}$, $\bar{\beta}$, and $\bar{\gamma}$ directions, respectively, and the angular displacements φ and ψ defined around the $\bar{\alpha}$ and $\bar{\beta}$ directions, respectively. The radius of the middle shell surface is r, the section thickness is h, the shear stiffness correction factor of the section is κ, the moment of inertia is J, the length is l, the elastic modulus of the material is E, the shear modulus is G, Poisson's ratio is μ, and density is ρ.

The governing differential equations of the free vibration of the shell are

$$(1/2)K\left[(1-\mu)n^2 u/r^2 - 2u'' - (1+\mu)nv'/r - 2\mu w'/r\right]$$
$$= \omega^2 \rho h u, \tag{9.1a}$$

$$(1/2)K\left[(1+\mu)nu'/r + 2n^2 v/r^2 - (1-\mu)(1+h^2/12r^2)v'' + 2nw/r^2\right.$$
$$\left. + (1-\mu)nh^2\phi'/(12r^2) - (1-\mu)h^2\psi''/(12r)\right]$$
$$+ \bar{\kappa}Gh(v/r^2 + nw/r^2 - \psi/r) = \omega^2 \rho h v, \tag{9.1b}$$

$$K(\mu u'/r + nv/r^2 + w/r) + \bar{\kappa}Gh(n^2 w/r^2 + nv/r^2 - n\psi/r - w'' - \phi')$$
$$= \omega^2 \rho h w, \tag{9.1c}$$

$$(h^2/24)K\left[-(1-\mu)nv'/r^2 + (1-\mu)n^2\phi/r^2 - 2\phi'' - (1+\mu)n\psi'/r\right]$$
$$+ \kappa Gh(w' + \phi) = \omega^2 \rho J \phi, \tag{9.1d}$$

$$(h^2/24)K\left[-(1-\mu)v''/r + (1+\mu)n\phi'/r + 2n^2\psi/r^2 - (1-\mu)\psi''\right]$$
$$+ \kappa Gh(\psi - v/r - nw/r) = \omega^2 \rho J \psi. \tag{9.1e}$$

where the prime mark ($'$) denotes the derivative with respect to the independent variable x, $K = Eh/(1-\mu^2)$ is the shell stiffness, ω is the frequency, and n is the circumferential wave number.

The solution idea of this study is not to write the eigensolutions of Equation (9.1) in explicit expression according to x and n for the five vibration mode components (which involves solving the complex partial differential equations of two variables). Instead, the value of n is given and determined first, and then the Equation (9.1) containing only variable x used to be solved (which involves solving simple ordinary differential equations with only one variable). The subsequent eigenproblem solved in this study includes determining the eigenvalues λ and associated n_d dimensional vector eigenfunctions $\boldsymbol{u}(x) = (u_1(x), \cdots, u_{n_d}(x))^{\mathrm{T}} = (u, v, w, \phi, \psi)^{\mathrm{T}}$ for the following system of second-order ordinary differential equations (ODEs) [59, 60]

$$\boldsymbol{Lu} \equiv \boldsymbol{Au}'' + \boldsymbol{Bu}' + \boldsymbol{Cu} = \lambda \boldsymbol{Ru}, \quad a < x < b, \tag{9.2}$$

where $\boldsymbol{L}$ is the differential operator, $\boldsymbol{A}$, $\boldsymbol{B}$, $\boldsymbol{C}$, and $\boldsymbol{R}$ are continuous $n_d \times n_d$ matrix functions on (a, b), and $n_d = 5$ is associated with the governing differential equations for the free vibration of the shell, as shown in Equation (9.1). Structural vibration problems were chosen to illustrate the possible physical interpretations of the equations. The corresponding structural natural frequencies ($\omega = \sqrt{\lambda}$) and modes represent the eigenvalues and vector eigenfunctions, respectively.

9.3   Damage characterisation method for circumferential cracks in circular cylindrical shell

In this study, the damage defect characterisation and rotating spring techniques used to characterise the micro-crack in the beams[58] are extended to this method to implement damage characterisation in a cylindrical shell. As shown in Figure 9.2, the geometric model of the moderately thick circular cylindrical shell embedded a circumferential crack, in which the parameters α and β denote the normalised crack depth and location, respectively:

$$\alpha = a/h, \tag{9.3a}$$
$$\beta = s/l, \tag{9.3b}$$

where a and s are the absolute crack depth and location, respectively, and h is the height of the shell.

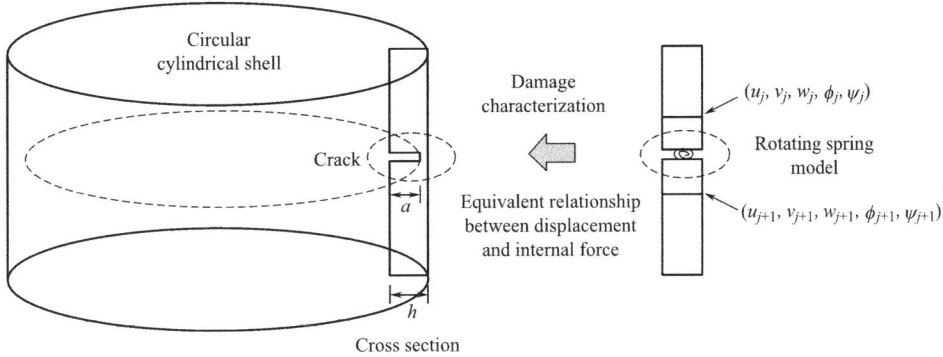

Figure 9.2   Damage characterisation for circumferential crack in circular cylindrical shell

The local domain around the micro-crack of the shell is shown in Figure 9.2. In the immediate region surrounding a single crack, the finite element containing the crack has two nodes with the degrees of linear and angular freedoms $(u_j, v_j, w_j, \phi_j, \psi_j)$ and $(u_{j+1}, v_{j+1}, w_{j+1}, \phi_{j+1}, \psi_{j+1})$, where the narrow crack is described with a width δ_c set at $0.01 \times Tol$ (Tol is the pre-specified error tolerance for both frequencies and modes). Using the weakened properties analogy to reflect the presence of cracks, the shell stiffness and moment of inertia at the crack are reduced as the crack deepens:

$$K_c = Eh(1-\alpha)/(1-\mu^2), \tag{9.4a}$$
$$J_c = \frac{bh^3(1-\alpha)^3}{12}, \tag{9.4b}$$

where EI_c and J_c are the shell stiffness and moment of inertia at the crack c, respectively, and b is the width of the shell (the value is taken as 1 representing the unit width).

9. 4 *h*-version mesh refinement method for eigensolutions of cracked circular cylindrical shell

9. 4. 1 Finite element solutions

The Lagrange interpolation formula provides a simple construct forhigher-order shape functions that satisfy Equation (9. 2).

$$l_a^p(\xi) = \prod_{\substack{b=1 \\ b \neq a}}^{n} \frac{(\xi - \xi_b)}{(\xi_a - \xi_b)} = \frac{(\xi - \xi_1)(\xi - \xi_2)(\cdots)(\xi - \xi_{a-1})(\xi - \xi_{a+1})(\cdots)(\xi - \xi_n)}{(\xi_a - \xi_1)(\xi_a - \xi_2)(\cdots)(\xi_a - \xi_{a-1})(\xi_a - \xi_{a+1})(\cdots)(\xi_a - \xi_n)}.$$

(9. 5)

The order of this polynomial is $p = n - 1$. Having chosen the end-node locations, the internal values of ξ_a may be spaced in uniform increments. For one-dimensional elements, we can set[61]

$$N_a(\xi) = l_a^p(\xi) \tag{9. 6}$$

to define the shape functions. The shape functions in finite element model used in this study are conventional shape functions, which makes the traditional finite element method can be put into use.

The weak form for the eigenproblems in the system ofsecond-order ODEs, as defined in Equation (9. 2), can be expressed as

$$\int_{\Omega} \{v'^{\mathrm{T}} A u' + v^{\mathrm{T}} [B u' + (C - \lambda R) u]\} \mathrm{d}x = 0, \tag{9. 7}$$

where v is a trial function and Ω is the solution domain. The finite element model uses the conventional degree m of polynomial elements. We let e denote a typical element with end-node coordinates $\overline{x}_1$ and $\overline{x}_2$ and with length h. We write the trial function on an element of degree m as

$$v = \sum_{i=1}^{m+1} N_i v_i, \tag{9. 8}$$

where N_i is an $n_d \times n_d$ shape function matrix defined by

$$N_i = N_i I, \quad i = 1, 2, \cdots, m+1, \tag{9. 9}$$

where I is the $n_d \times n_d$ identity matrix.

Using the conventionalfinite element method, the element stiffness and mass matrices (K^e and M^e, respectively) are computed and assembled to form the global stiffness and mass matrices K and M, respectively. The finite element equation is then derived as an eigenvalue equation in the following matrix form:

$$KD = \lambda MD, \tag{9. 10}$$

where D is the eigenfunction vector, and matrices K and M are independent of λ. Based on the inverse iteration technique[58], the eigensolutions of Equation (9. 10) can be derived.

9. 4. 2   Error estimation

The superconvergent patch recovery displacement method was developed[56-58, 42] to acquire the superconvergent displacements of the finite element solutions in static and dynamic problems. The displacements provided by this method can be applied to eigenfunctions. For example, if element e is a superconvergent computation element and elements $e-1$ and $e+1$ are its neighbouring elements, all finite element nodes in patched elements $e-1$, e, and $e+1$ are selected for the computation process. Further, the superconvergent displacements for element e can be computed as

$$\boldsymbol{u}^*(x) = \sum_{i=1}^{r} \mathbf{N}_i(x)\boldsymbol{u}_i^h + \sum_{i=1}^{s} \mathbf{N}_i(x)\overline{\boldsymbol{u}}_i^*, \qquad (9.11)$$

where $r=2$ is the number of end nodes, s is the number of internal nodes, and $\mathbf{N}_i$ is the shape function matrix. Using high-order shape function interpolation, the polynomial order of the shape function is increased, $r+s>m+1$. To optimise the superconvergent order $O(h^{2m})$ for displacements at the end nodes, the displacement recovery field can be expressed for the finite element nodes as follows:

$$\overline{u}_i^*(x) = \boldsymbol{P}\boldsymbol{a}, \quad i=1, \cdots, n_d, \qquad (9.12)$$

where $\boldsymbol{P}$ is the given function vector and $\boldsymbol{a}$ can be determined by least-squares fitting for the coincidence of displacements at the end nodes in both the recovery and conventional finite element fields. The superconvergent displacements of the recovery field are used in Equation (9.11) to obtain the superconvergent solutions of element e. We use the following forms for the vector coefficients $\boldsymbol{P}$ and $\boldsymbol{a}$:

$$\boldsymbol{P} = \begin{bmatrix} 1 & x & \cdots & x^p \end{bmatrix}, \quad \boldsymbol{a} = \begin{bmatrix} a_1 & a_2 & \cdots & a_m \end{bmatrix}^{\mathrm{T}} \qquad (9.13)$$

The value of $\boldsymbol{a}$ was determined from the minimum value of the following functional, so that the product of $\boldsymbol{P}$ and $\boldsymbol{a}$, computed using Equation (9.12), matches the displacement values:

$$\Pi = \sum_{j=1}^{n_p} (u_i^*(x_j) - \boldsymbol{P}(x_j)\boldsymbol{a}), \quad i=1, \cdots, n_d, \qquad (9.14)$$

where n_p is the node number of all elements patched together.

The least-squares method applied to Equation (9.14) yields

$$\boldsymbol{a} = \boldsymbol{A}^{-1}\boldsymbol{b}, \qquad (9.15)$$

with the coefficient matrices $\boldsymbol{A}$ and $\boldsymbol{b}$ defined as:

$$\boldsymbol{A} = \sum_{j=1}^{n} \boldsymbol{P}(x_j)^{\mathrm{T}}\boldsymbol{P}(x_j), \quad \boldsymbol{b} = \sum_{j=1}^{n} \boldsymbol{P}(x_j)^{\mathrm{T}}u_i^h(x_j), \quad i=1, \cdots n_d. \qquad (9.16)$$

After $\boldsymbol{a}$ is determined, the superconvergent solutions of the displacement of the piecewise elements are obtained from Equation (9.11). The estimated eigenvalue has a stationary value when all possible functions satisfying the essential boundary conditions are considered. These stationary values are the superconvergent eigenvalues. Furthermore, the superconvergent solutions of the displacements can be used in the Rayleigh quotient[62, 63] to estimate the eigenvalue as

$$\lambda^* = \frac{a(\boldsymbol{u}^*, \boldsymbol{u}^*)}{b(\boldsymbol{u}^*, \boldsymbol{u}^*)}, \tag{9.17}$$

where $a(\cdot)$ and $b(\cdot)$ are the strain and kinematic energy inner products, respectively. To determine whether the solution for the considered mesh meets the required tolerance condition, the error must satisfy the following condition:

$$\|\boldsymbol{e}^*\| \leqslant Tol \cdot [(\|\boldsymbol{u}^h\|^2 + \|\boldsymbol{e}^*\|^2)/n_e]^{1/2}, \tag{9.18}$$

where n_e is the number of elements, and $\|\boldsymbol{e}^*\|$ is the error in the energy norm, as follows:

$$\|\boldsymbol{e}^*\| = [a(\boldsymbol{e}^*, \boldsymbol{e}^*)]^{1/2} = \left[\int_\Omega \boldsymbol{e}^{*\mathrm{T}} \boldsymbol{L} \boldsymbol{e}^* \, \mathrm{d}x\right]^{1/2}, \tag{9.19}$$

where $\boldsymbol{e}^* = \boldsymbol{u}^* - \boldsymbol{u}^h$, and Ω is the solution domain.

9.4.3 Element subdivision and refinement

Equation (9.18) can be rewritten as

$$\xi \leqslant 1, \tag{9.20}$$

where

$$\xi = \frac{\|\boldsymbol{e}^*\|}{\overline{e}} \quad \text{with} \quad \overline{e} = Tol \cdot [(\|\boldsymbol{u}^h\|^2 + \|\boldsymbol{e}^*\|^2)/n_e]^{1/2}. \tag{9.21}$$

If Equation (9.20) is not satisfied, the corresponding element is subdivided into identical sub-elements by inserting interior nodes through h-refinement[58, 42]. These are computed using

$$h_{new} = \xi^{-1/m} h_{old}, \tag{9.22}$$

where h_{new} is the length of the sub-element and h_{old} is the original length of element e. The above element subdivision approach was implemented as follows:

$$n_{new} = \min(\lfloor \xi^{1/m} \rfloor, d), \tag{9.23}$$

where n_{new} is the number of sub-elements after element subdivision, the symbol $\lfloor \cdot \rfloor$ denotes the "floor" operator (i.e., the rounding down to the nearest integer), and d is the limit needed to avoid too many redundant elements. Each element e that does not meet the pre-specified error tolerance threshold is uniformly subdivided by the h-version mesh refinement.

9.5 Global algorithm and procedure

According to the finite element solution, error estimation, mesh subdivision, and refinement methods for free vibration of cylindrical shells with crack damage introduced above, the adaptivefinite element algorithm for free vibration disturbance of moderately thick circular cylindrical shells with circumferential micro-crack damage can be established, as shown in Figure 9.3. To briefly illustrate the implementation of the above formula in the program, the pseudo code algorithm for eigensolutions of cylindrical shells with circumferential micro-crack damage is shown in Algorithm 9.1. First, the basic parameters including the micro-crack damage, physical parameters, circumferential wave numbers, initial mesh, and error *tolerance* should be provided.

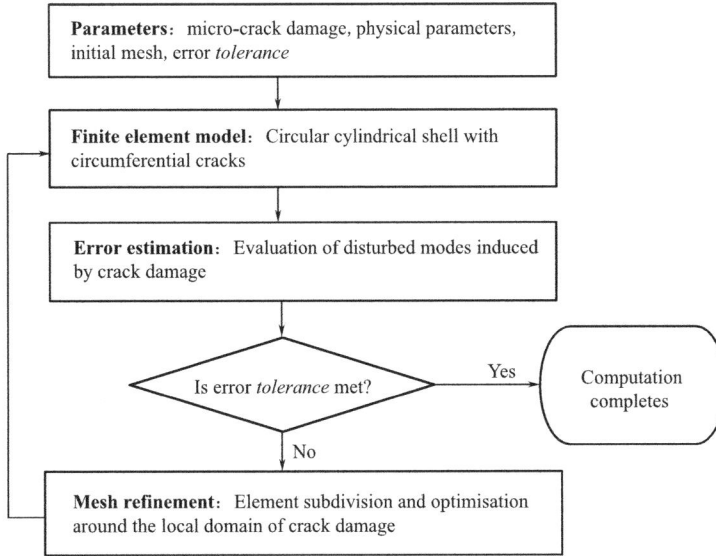

Figure 9.3 Flow chart of the adaptive finite element algorithm and procedure for free vibration disturbance of moderately thick circular cylindrical shells with circumferential micro-crack damage

The procedure involves the following three basic processes:

(1) **Finite element model:** Based on the circular cylindrical shell with circumferential cracks and damage characterisation method, the conventional finite element computation is performed based on the initial mesh. The finite element solutions (ω^h, $\boldsymbol{u}^h$) under the current mesh were computed.

(2) **Error estimation:** Based on the error estimation technique, the disturbance modes induced by crack damage can be evaluated. It should be noted that in the vicinity of crack damage, owing to the disturbance of the vibration mode, its solution may produce significant errors. At the same time, the information on whether the error tolerance of each element is met is marked.

(3) **Mesh refinement:** Element subdivision and optimisation around the local domain of crack damage. For the element whose error estimation does not meet the error tolerance condition, the mesh subdivision refinement method is used to subdivide it and obtain a new finite element mesh before returning to step (1). In the local domain of crack damage, owing to the large error, it may be necessary to introduce more subdivided elements. If all the elements meet the error tolerance condition, no subdivision is necessary, and the computation process is completed. Through the above adaptive process, the optimal mesh and high-precision solution under the current eigenvalue order of the cylindrical shell with

crack damage can be obtained when the vibration mode is disturbed, which is especially suitable for cylindrical shells with crack damage when the vibration mode is disturbed.

Pseudo code algorithm for eigensolutions of cylindrical shells with

circumferential micro-crack damage　　　　　　　　　**Algorithm 9. 1**

♯ **Input:** Micro-crack damage (number n_c, depth α, and location β), physical parameters, circumferential wave number n, initial mesh π_0, error *tolerance Tol*

♯ **Finite element model:**

$K_c = Eh(1-\alpha)/(1-\mu^2)$　　! Parameters at cracks

$J_c = \dfrac{bh^3(1-\alpha)^3}{12}$

do $k = 1$, k_n　　　　　　　　! k is the current order of eigenvalue

　　KD $= \lambda$**MD** $\to \omega$, D　　! Inverse power iteration for eigensolutions on π

　　♯ **Error estimation:**

　　$\bar{u}_i^*(x) = $ **Pa**　　　　　! Superconvergent patch recovery displacement

　　do $\bar{n} = 1$, n_e　　　　　! Check and mark error element by element

　　　　$\xi \leqslant 1$, $\xi = \dfrac{\| \mathbf{e}^* \|}{\bar{e}}$ with $\bar{e} = Tol \cdot [(\| \mathbf{u}^h \|^2 + \| \mathbf{e}^* \|^2)/n_e] 1/2$　　! Error check

　　end do

　　♯ **Mesh refinement:**

　　$h_{\text{new}} = \xi^{-1/m} h_{\text{old}}$　　　　! Subdivision for the element with large error

　　$\pi = \pi + 1$　　　　　　　　! Update mesh

　　♯ **Output:** Eigensolutions (ω, D)

end do

---

9. 6　Numerical examples

This section presents some representative numerical examples to demonstrate the performance of the proposed method and algorithm, which were implemented in Fortran 90. The uncracked and cracked damage cases of moderately thick circular cylindrical shells were computed, and the reliability and accuracy of the method described herein were verified. The effects of different crack locations, crack depths, and number of multiple cracks on the free vibration disturbance were analysed. The program was run on a DELL Optiplex 380 Intel® Core™ 2. 93 GHz desktop computer. The degree of element for mesh refinement is $m = 3$, the initial number of elements used in the computation procedure is $n_e = 2$, and the stricter pre-specified error tolerance is $Tol = 10^{-6}$. In this study, the mode error introduced in Section 4. 2 was used for solution estimation and control. The frequency was computed by the Rayleigh quotient, the error of which needs to be estimated. The relative error ε_ω between the exact frequency $\bar{\omega}$ (i. e. , the high-precision solutions from other methods) and the computed frequency ω^h was defined and used to analyse the precision of the solutions:

$$\varepsilon_\omega = \frac{|\bar{\omega} - \omega^h|}{1 + |\bar{\omega}|} \tag{9. 24}$$

9. 6. 1   Example 1: Benchmarks for free vibration of circular cylindrical shell

To verify the reliability and accuracy of the methodproposed in this study, the free vibration frequency of a circular cylindrical shell without crack damage was computed under the condition of simple support at both ends. The basic geometric and physical parameters of the cylindrical shell are listed in Table 9. 1.

Geometric and physical parameters of shell in Example 1   Table 9. 1

Parameters	r (mm)	h (mm)	l (mm)	μ	κ	E (GPa)	ρ (kg $\cdot$ m^{-3})
Values	148. 234	0. 508	298. 2	0. 285	5/6	203. 5	7846

The method presented in thisstudy was used to solve the first-order ($k=1$) solutions with different circumferential wave numbers n ($n=1-5$). The computed frequency results are presented in Table 9. 2. In previous studies (Chen and Ye, 2016; Sivadas and Ganesan, 1994), the dynamic stiffness method and the hybrid finite element method were used to solve the problem, respectively, based on the medium-thick shell theory. It can be observed that the results in this study are in good agreement with those frequency solutions, which verifies the reliability of the method in solving the frequency of each order.

Computed frequencies ω/Hz of shell in Example 1   Table 9. 2

n	ω_1^h	$\omega_1^{h\,①}$	$\omega_1^{h\,②}$
1	3270. 74	3270. 6	3270. 9
2	1862. 12	1862. 0	1862. 1
3	1101. 83	1101. 8	1101. 8
4	705. 758	705. 9	705. 7
5	497. 467	497. 5	497. 4

Source: ①Results from paper [26]; ②Results from paper [6].

Figure 9. 4 shows the typical computed vibration modes and the corresponding final meshes on the horizontal x-axis for circumferential wave numbers $n=1$. The components (u^h, v^h, w^h) and (ϕ^h, ψ^h) of the first-order vibration mode with approaching magnitude are shown in Figures 4 (a) and 4 (b), respectively. It should be noted that to facilitate visual display and comparative analysis, the vibration mode results in this study were normalised (make the maximum vibration mode value 1), and the horizontal x-axis is also normalised in the vibration mode diagram (the horizontal coordinate axis x/l). It can be observed that the vibration mode changes sharply at both ends, and the relatively fine mesh is divided by the adaptive refinement method. The variation of the vibration mode is relatively gentle in the middle domain, and only sparse meshes are optimised and provided. The computation results of the medium-thick cylindrical shell indicate that the adaptive strategy herein is accurate and reliable, and the mesh division is reflected in the error analysis and mesh division described above.

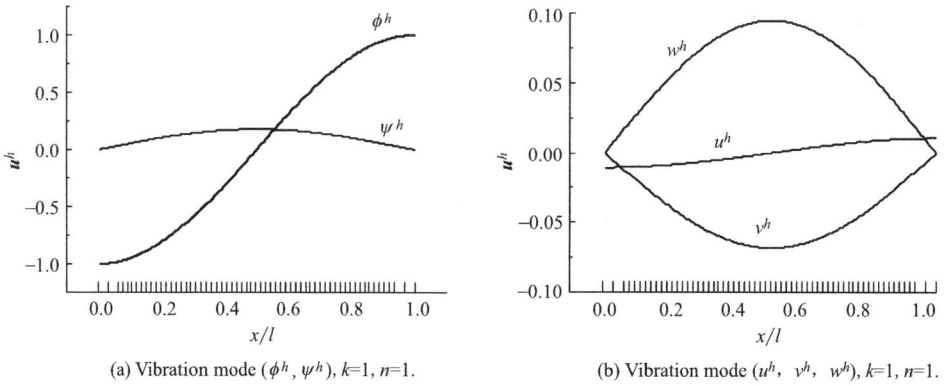

(a) Vibration mode (ϕ^h, ψ^h), k=1, n=1.　　　(b) Vibration mode (u^h, v^h, w^h), k=1, n=1.

Figure 9.4　Example 1: computed vibration modes and corresponding final meshes for circumferential wave numbers $n=1$

9.6.2　Example 2: Verification of frequency solutions of cracked circular cylindrical shell under variable circumferential wave numbers

The effectiveness of the method proposed in this study for the analysis of cylindrical shells with crack damage was tested. The shell had a crack damage located at $\beta=0.3$ with a depth $\alpha=0.6$. The solutions under different frequency orders and circumferential wave numbers were computed and discussed. The geometric and physical parameters of the shell are listed in Table 9.3.

Geometric and physical parameters of shell in Example 2　　　　**Table 9.3**

Parameters	r (mm)	h (mm)	l (mm)	μ	κ	E (GPa)	ρ (kg · m^{-3})
Values	100	2	500	0.3	5/6	210	7850

Using the proposed method, the frequencies and vibration modes of cracked cylindrical shells with different orders k ($k=1-3$) and circumferential wave numbers n ($n=1-5$) were solved. Table 9.4 lists the computed frequency solutions under various circumferential wave numbers. To solve the numerical example of the free vibration of a cylindrical shell with damage, the results were obtained using the combined scheme constructed by conventional finite element method and the beam function and Soedel's expression method[22]. For comparative analysis, the relative error between the frequency results of the proposed method and the frequency result computed in the above study are listed in Table 9.4, which shows that the error is very small, verifying the effectiveness of the method in this study. At the same time, the results of the final adaptive mesh number n_e for various cases are presented.

To illustrate the disturbance of the vibration mode caused by crack damage and the adaptive subdivisionand refinement behaviour of the mesh, Figure 9.5 shows the representative computed vibration modes and corresponding final meshes for a circular wave number $n=3$.

188

Computed frequencies of cracked circular cylindrical shell
under variable circumferential wave numbers in Example 2　　Table 9. 4

k	n	Conventional finite element method ω^h [1]	Beam function and Soedel's expression method ω^h [1]	ω^h	Error (%)	n_e
1	3	509. 10	510. 54	511. 47	0. 18	85
1	2	665. 15	652. 61	658. 32	0. 87	93
1	4	766. 54	767. 28	769. 66	0. 31	88
2	4	1061. 98	1066. 42	1065. 26	0. 11	110
1	5	1199. 35	1196. 21	1197. 84	0. 13	91
2	3	1229. 20	1216. 05	1221. 25	0. 43	111
2	5	1332. 13	1335. 81	1338. 43	0. 20	113
1	1	1598. 83	1566. 33	1548. 92	1. 11	106
3	5	1678. 17	1674. 43	1678. 04	0. 22	157
3	4	1703. 03	1696. 90	1687. 82	0. 53	161

Source: [1]Results from paper [22].

It can be observed that in the domain around the crack damage, the vibration mode changes, and there is a large deformation and disturbance. In the domain around the crack, the adaptive procedure generates a very dense mesh to capture more refined vibration mode changes.

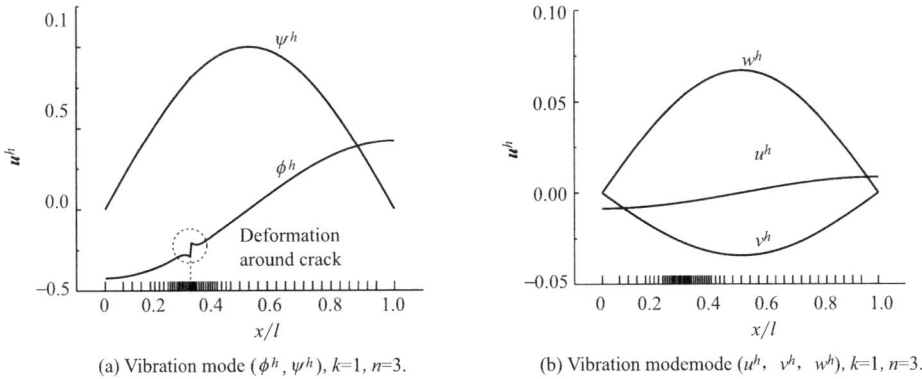

(a) Vibration mode (ϕ^h, ψ^h), k=1, n=3.　　(b) Vibration modemode (u^h, v^h, w^h), k=1, n=3.

Figure 9. 5　Example 2: computed vibration modes and corresponding
final meshes for circumferential wave number $n=3$

To show the disturbance effect ofcircumferential crack damage and circumferential wave number on spatial vibration modes, Figure 9. 6 shows the spatial morphologies of the vibration modes for a circular wave number $n=3$. When the circumferential wave number n is 3, there are multiple circumferential waves in the circumferential direction among the five vibration mode components. In the spatial local domain of the circumferential crack damage, the vibration mode fluctuates, which is most obvious in the components shown in Figure 9. 6 (d). It can be seen that for the identical crack damage in Equation (9. 4), the rotation stiffness decreases more than the bending stiffness, which will induce greater disturbance of rotation deformation. The crack damage in this study is very tiny to describe the micro-crack, if the crack damage continues to increase (such as increasing the crack width), it will bring significant disturbance to other vibration mode components. A non-uniform mesh was generated on the horizontal axis. According to the above analysis, dense meshes were used in the domain around the crack to solve the five vibration mode components simultaneously.

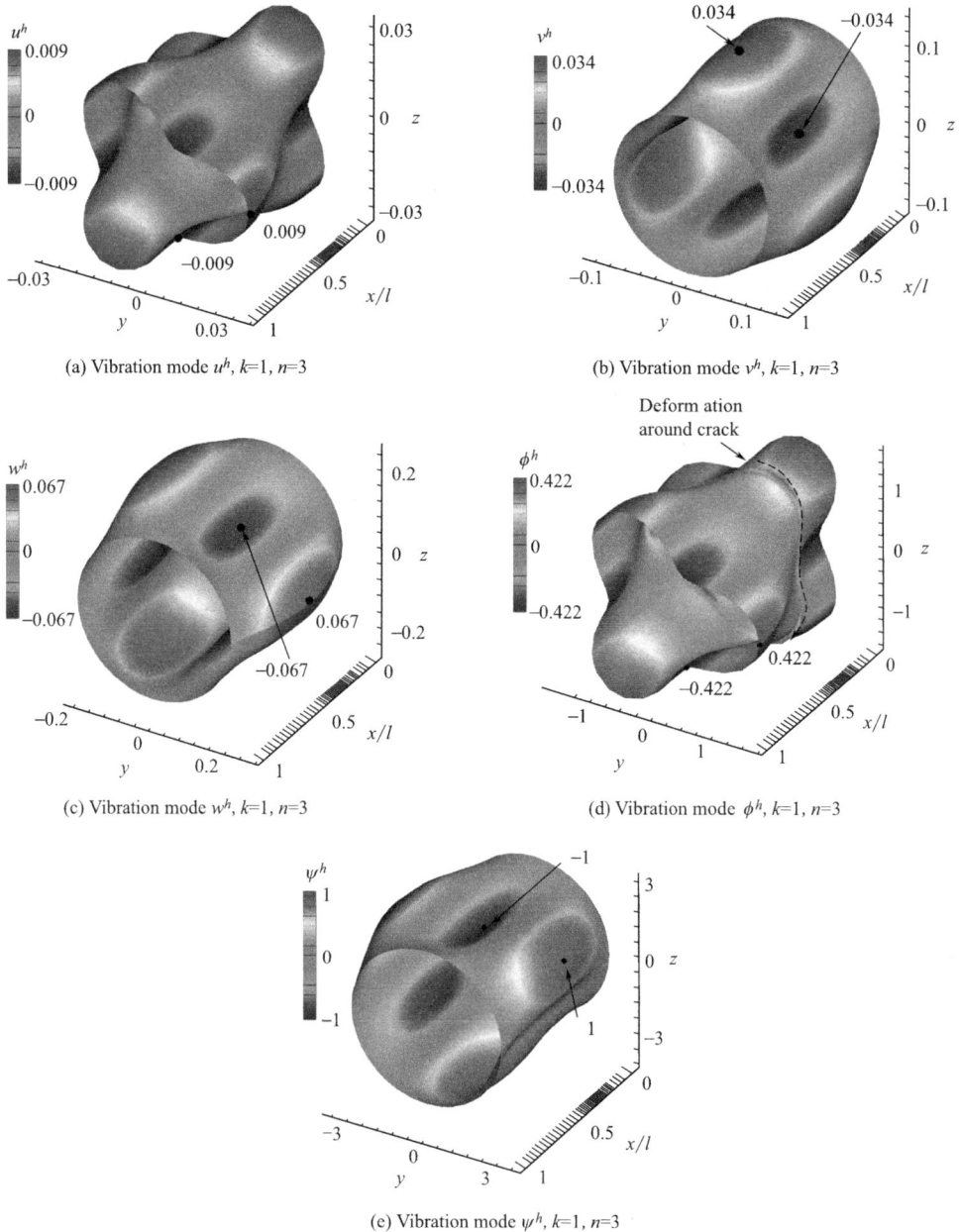

(a) Vibration mode u^h, $k=1$, $n=3$

(b) Vibration mode v^h, $k=1$, $n=3$

(c) Vibration mode w^h, $k=1$, $n=3$

(d) Vibration mode ϕ^h, $k=1$, $n=3$

(e) Vibration mode ψ^h, $k=1$, $n=3$

Figure 9.6　Example 2: Spatial morphologies of vibration modes for circumferential wave number $n=3$

To show the influence of different circumferential wave numbers and vibration orders on vibration modes and meshing, Figure 9.7 shows the representative results of the computed vibration modes and corresponding final meshes for different order k of eigenvalue and circular wave numbers n. It can be observed from Figures 7 (a)~(d) that with the increase in the wave number of the circumferential vibration mode, the change in the vibration mode along the axial direction is not obvious, but the change in the circumferential vibration mode will show the increase in the wave number. It can be observed from Figures 7 (e)~(f) that with the increase in vibration order, the change in

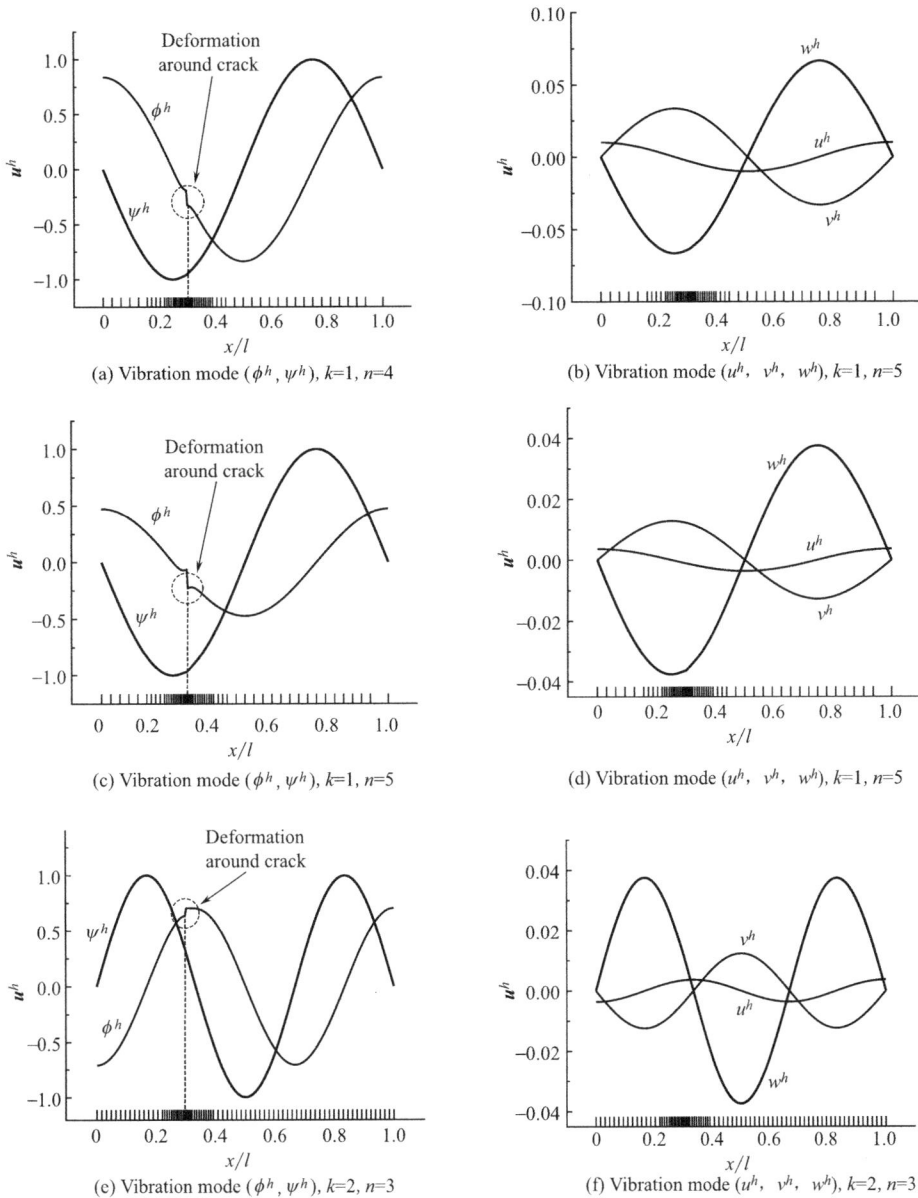

(a) Vibration mode (ϕ^h, ψ^h), $k=1$, $n=4$

(b) Vibration mode (u^h, v^h, w^h), $k=1$, $n=5$

(c) Vibration mode (ϕ^h, ψ^h), $k=1$, $n=5$

(d) Vibration mode (u^h, v^h, w^h), $k=1$, $n=5$

(e) Vibration mode (ϕ^h, ψ^h), $k=2$, $n=3$

(f) Vibration mode (u^h, v^h, w^h), $k=2$, $n=3$

Figure 9.7　Example 2: computed vibration modes and corresponding final meshes for different circumferential wave numbers

vibration mode along the axial direction will become more complex and more circumferential wave numbers in the vibration modes will appear, which also makes the number of elements in the whole domain greater than in the first-order vibration mode described in Figures 7 (a)~(d). It can be concluded that the circumferential wave number of rotating shells (such as the circular cylindrical shell in this study) will not have a significant difference with respect to the mesh distribution owing to the symmetry of the vibration mode, but the change in the order of the vibration mode puts forward higher requirements for the number of elements in the whole region. The existence of crack damage makes it necessary to refine the local mesh to capture the disturbance change in the vibration mode.

9. 6. 3 Example 3: Free vibration disturbance by crack location

This example discusses the influence of the crack damage location on the vibration disturbance of a cylindrical shell. This example uses the model parameters of Example 2 to change the crack damage location. The crack location β was changed from 0. 1 to 0. 9 with an interval of 0. 1. Table 9. 5 lists the computed first-order frequencies for different crack locations. Here, the number of final adaptive elements n_e used is also provided. Figure 9. 8 shows the relationship between the computed frequency and crack location to show the frequency values at different crack damage locations. As can be observed, when the crack damage is at a symmetrical location (such as $\beta=0. 2$ and 0. 8), the same frequency value will appear owing to the structural symmetry of the shell. As indicated in Table 9. 5, the same number of elements was used. At the same time, as shown in Figure 9. 8, when the crack damage occurred at both ends of the cylindrical shell, the stiffness of the shell decreased the most, resulting in a relatively low frequency value. At this time, the stiffness of the shell changed significantly, and more elements were used than when the crack was located in the middle of the shell.

Computed frequencies under different crack locations in Example 3   **Table 9. 5**

β	ω^h	n_e
0. 1	10107. 48	106
0. 2	10112. 95	112
0. 3	10119. 72	106
0. 4	10125. 22	96
0. 5	10127. 32	88
0. 6	10125. 22	96
0. 7	10119. 72	106
0. 8	10112. 95	112
0. 9	10107. 48	106

Figure 9.9 shows the computed vibration modes and corresponding final meshes for different crack locations $\beta=0.1$ and $\beta=0.5$, respectively. It can be observed that the location of crack damage causes a corresponding disturbance to the vibration mode components, and the local subdivision and refinement domain of the mesh changes with the change in the vibration mode location of the crack damage disturbance. These results demonstrate the effectiveness of the adaptive mesh refinement method on free vibration disturbance by crack location.

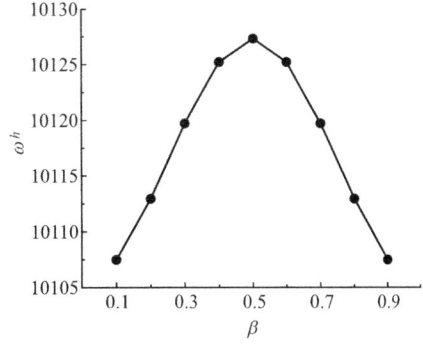

Figure 9.8   Example 2: Relationship between computed frequency and crack location

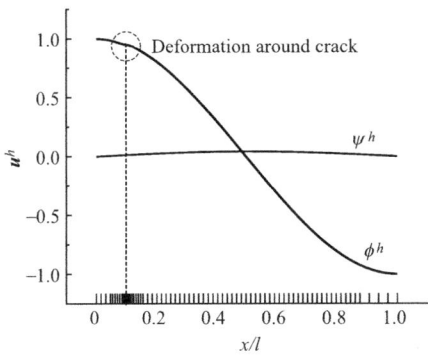

(a) Vibration mode (ϕ^h, ψ^h), $k=1$, $n=1$, $\beta=0.1$

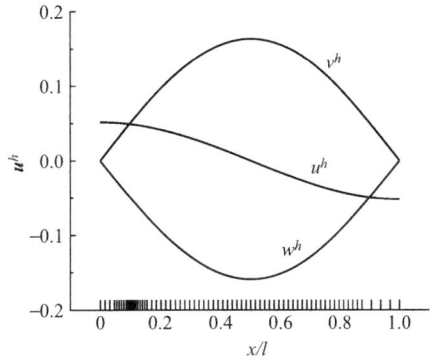

(b) Vibration mode (u^h, v^h, w^h), $k=1$, $n=1$, $\beta=0.1$

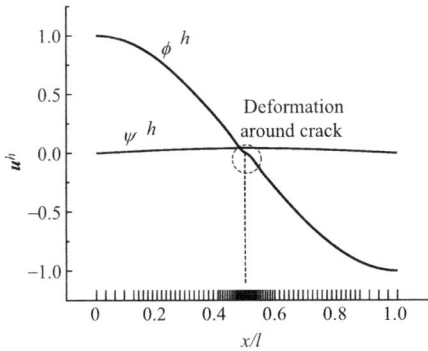

(c) Vibration mode (ϕ^h, ψ^h), $k=1$, $n=1$, $\beta=0.5$

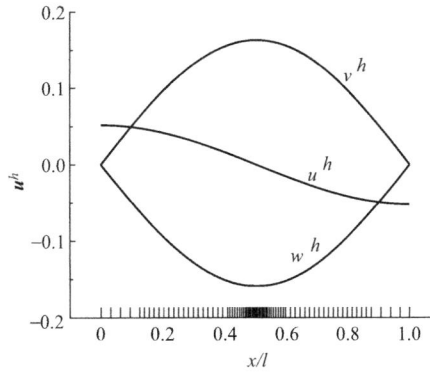

(d) Vibration mode (u^h, v^h, w^h), $k=1$, $n=1$, $\beta=0.5$

Figure 9.9   Example 3: Computed vibration modes and corresponding final meshes for different crack locations

9.6.4   Example 4: Free vibration disturbance by crack depth

This example discusses the influence of the crack damagedepth on the vibration disturbance of a cylindrical shell. This example uses the model parameters of Example 2 to

193

change the crack damage depth. The crack depth α was changed from 0.2 to 0.6 with an interval of 0.2. Table 9.6 lists the computed first-order frequencies for different crack depths. Here, the number of final adaptive elements n_e used is also provided. It can be observed that the change in crack damage depth changes the frequency slightly, but the increase in the degree of crack damage increases the vibration mode disturbance and increases the number of required elements n_e.

Computed frequencies under different crack depths in Example 4 Table 9.6

α	ω^h	n_e
0.2	10149.79	88
0.4	10140.20	93
0.6	10119.72	106

Figure 9.10 shows the computed vibration modes and corresponding final meshes for different crack depths $\alpha=0.2$ and $\alpha=0.4$, respectively. As can be observed, the depth of crack damage causes a corresponding disturbance of the vibration mode component, but the entire shape is basically the same as that in the different crack damage cases. The mesh distribution is basically the same in the entire domain, but it changes in the local domain of

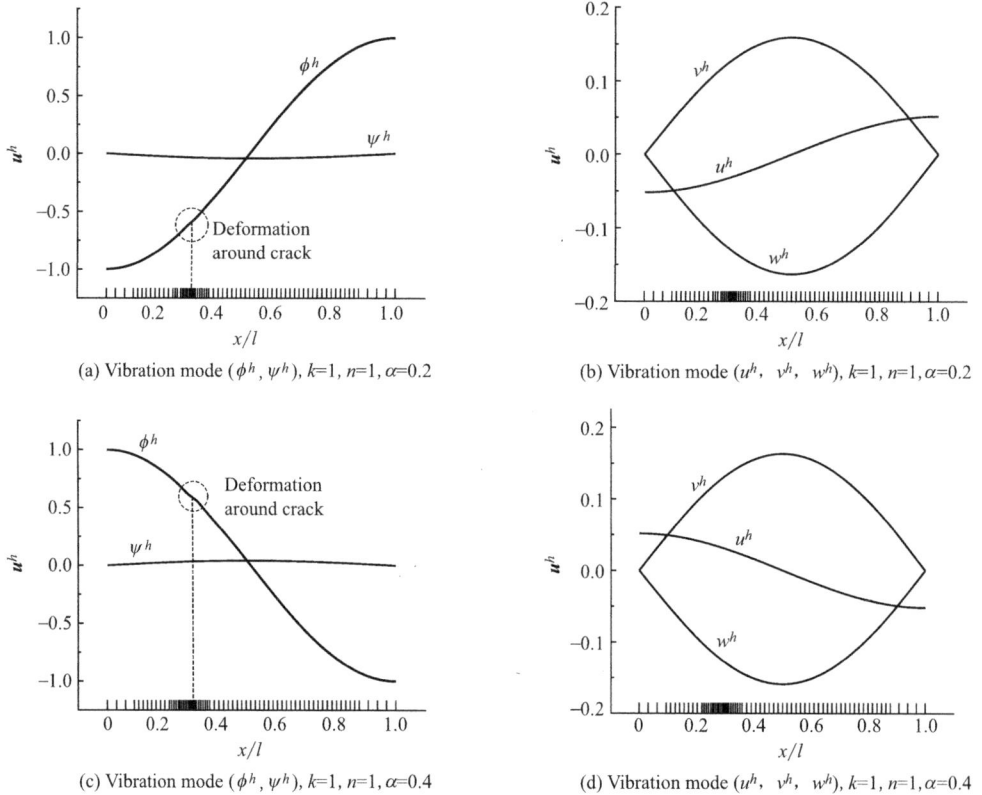

(a) Vibration mode (ϕ^h, ψ^h), $k=1$, $n=1$, $\alpha=0.2$

(b) Vibration mode (u^h, v^h, w^h), $k=1$, $n=1$, $\alpha=0.2$

(c) Vibration mode (ϕ^h, ψ^h), $k=1$, $n=1$, $\alpha=0.4$

(d) Vibration mode (u^h, v^h, w^h), $k=1$, $n=1$, $\alpha=0.4$

Figure 9.10   Example 4: computed vibration modes and corresponding
final meshes for different crack depths

the crack damage disturbance mode. The deeper the crack damage is, the more severe the mode damage, and more elements are needed.

9.6.5　Example 5: Free vibration disturbance by number of multiple cracks

This example discusses the influence of multiple-crack damage on the vibration disturbance of a cylindrical shell. This example uses the model parameters of Example 2 to change the number and location of multiple cracks. The crack location β was 0.3, 0.5, 0.7 (for three cracks) and 0.1, 0.3, 0.5, 0.7, 0.9 (for five cracks), respectively. Table 9.7 lists the computed frequencies for multiple cracks. Here, the number of final adaptive elements n_e used is also provided. As can be observed, the change in the number and location of cracks changes the frequency slightly, but the increase in the number of cracks increases the vibration mode disturbance and increases the number of required elements n_e. Compared with 190 elements for the computation of three cracks in the shell, 276 elements are needed for five cracks in the shell.

	Computed frequencies under multiple cracks in Example 5		Table 9.7
n_c	β	ω_1^h	n_e
3	0.3, 0.5, 0.7	10030.31	190
5	0.1, 0.3, 0.5, 0.7, 0.9	10018.87	276

Figure 9.11 shows the computed vibration modes and corresponding final meshes for different numbers of multiple cracks. Figures 9.11 (a) and (b) depict the vibration mode components under three-crack damage, while Figures 9.11 (c) and (d) show the vibration mode components under five-crack damage. It can be observed that the number and location of multiple-crack damage cause a corresponding disturbance to the vibration mode components, and the local subdivision and refinement domains of the meh changes with the change in the vibration mode location of the crack damage disturbance. The existence of

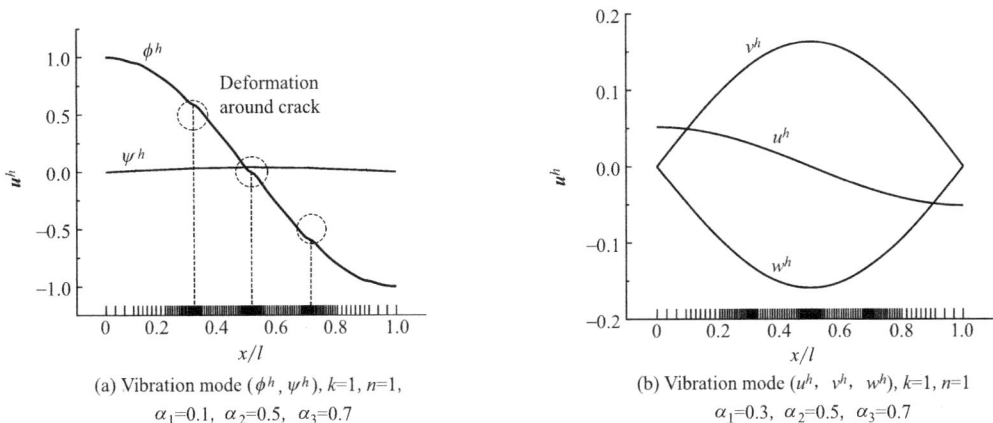

(a) Vibration mode (ϕ^h, ψ^h), $k=1$, $n=1$,
$\alpha_1=0.1$, $\alpha_2=0.5$, $\alpha_3=0.7$

(b) Vibration mode (u^h, v^h, w^h), $k=1$, $n=1$
$\alpha_1=0.3$, $\alpha_2=0.5$, $\alpha_3=0.7$

Figure 9.11　Example 5: computed vibration modes and corresponding
final meshes for different numbers of multiple cracks (one)

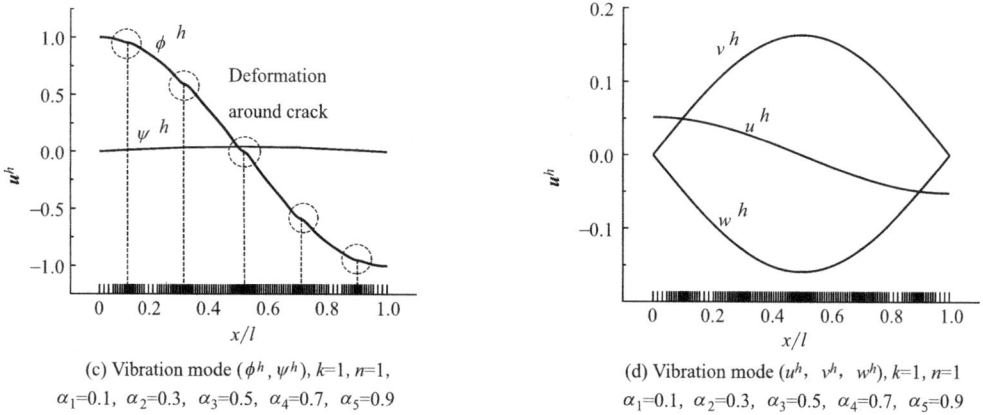

(c) Vibration mode (ϕ^h, ψ^h), $k=1$, $n=1$,
$\alpha_1=0.1$, $\alpha_2=0.3$, $\alpha_3=0.5$, $\alpha_4=0.7$, $\alpha_5=0.9$

(d) Vibration mode (u^h, v^h, w^h), $k=1$, $n=1$
$\alpha_1=0.1$, $\alpha_2=0.3$, $\alpha_3=0.5$, $\alpha_4=0.7$, $\alpha_5=0.9$

Figure 9. 11   Example 5: computed vibration modes and corresponding
final meshes for different numbers of multiple cracks (two)

multiple-crack damage causes multiple disturbances in the vibration modes of the shell, and hence more local domains need to be subdivided and densified, resulting in a large number of elements in the global domain. These results show the reliability of the adaptive mesh refinement method in this study to the changes in mode shapes and modes under simultaneous multiple-crack damage.

9. 7   Conclusions

In thisstudy, an adaptive mesh refinement analysis of the finite element method for free vibration disturbance of moderately thick circular cylindrical shells with circumferential crack damage is implemented to derive the frequency solutions under variable circumferential wave numbers and discuss the free vibration disturbance by factors of crack damage, such as the location, depth, and number of cracks. The conclusions of this study can be summarised as follows.

(1) An adaptive finite element method and crack damage characterisation method for moderately thick circular cylindrical shells are proposed. By introducing the inverse power iteration method, error estimation, and mesh subdivision refinement technique for the analysis of finite element eigenvalue problems, an adaptive computation scheme is constructed for the free vibration problem of moderately thick circular cylindrical shells with circumferential crack damage.

(2) Typical numerical examples of benchmarks and frequency solutions of cracked shells under variable circumferential wave numbers confirmed that the established adaptive finite element solution for the free vibration of moderately thick circular cylindrical shells is suitable for solving the high-precision free vibration frequency and mode of cylindrical shell structures. The reliability, accuracy, and effectiveness of the proposed method and models were verified.

196

(3) The adaptive mesh refinement algorithm has good applicability for the analysis of cracked shells. The vibration mode was disturbed near the crack damage. In thisstudy, the non-uniform mesh was adaptively optimised, and a relatively dense mesh was used near the crack to adapt to the change in vibration mode caused by crack damage.

(4) The occurrence of crack damage reduces the frequency of each order, and the greater the damage depth is, the greater the reduction degree. The crack damage generates the greatest disturbance to the rotational displacement, and the greater the damage degree is, the greater the disturbance amplitude. The number and location of cracks simultaneously affect the frequency value. An increase in the number of cracks tends to increase the frequency as a whole. However, because the change in crack location will also reduce the frequency value in some order frequencies, the vibration modes near each crack damage will be disturbed. Compared with the uniform distribution, the frequency value of the concentrated distribution on one side of the crack has a higher value at low order and a lower value at high order. The vibration modes were disturbed near the uniformly distributed and centrally distributed crack damage. The different distribution forms of the same number of cracks have become an important factor affecting the vibration characteristics.

This study can be used as a reference for the adaptive finite element solution of free vibration of moderately thick circular cylindrical shells with cracks and it lays a foundation for further development of a high-performance computation method suitable for dynamic disturbance and damage identification analysis of general cracked structures. Furthermore, the error estimation and element refinement techniques of the finite element method have the potential to be extended in the future to the refined numerical model and high-precision computation field of general structural eigenvalue problems (displacement field) and solid stress (displacement derivative field). Some studies on the validation and application of these adaptive methods are ongoing and will be reported in the future.

References

[1] Ide S, Yabe S, Tanaka Y, et al. Earthquake potential revealed by tidal influence on earthquake size-frequency statistics [J]. Nature Geoscience, 2016, 9 (11): 834-837.

[2] Chestler S R, Creager K C. Evidence for a scale-limited low-frequency earthquake source process [J]. Journal of Geophysical Research: Solid Earth, 2017, 122 (4): 3099-3114.

[3] Dey T, Ramachandra L S. Non-linear vibration analysis of laminated composite circular cylindrical shells [J]. Composite Structures, 2017, 163 (8).

[4] Arbelo M A, SFMD Almeida, Donadon M V, et al. Vibration correlation technique for the estimation of real boundary conditions and buckling load of unstiffened plates and cylindrical shells [J]. Thin-Walled Structures, 2014, 79 (1): 119-128.

[5] Kang B, Riedel C H, Tan C A, et al. Free vibration analysis of planar curved beams by wave propagation [J]. Journal of Sound and Vibration, 2003, 260 (1): 19-44.

［6］ Sivadas K R, Ganesan N. Free vibration and material damping analysis of moderately thick circular cylindrical shells ［J］. Journal of Sound & Vibration, 1994, 172 (1): 47-61.

［7］ Qu Y, Hua H, Meng G, et al. A domain decomposition approach for vibration analysis of isotropic and composite cylindrical shells with arbitrary boundaries ［J］. Composite Structures, 2013, 95: 307-321.

［8］ Weingarten V I. Free Vibration of Thin Cylindrical Shells ［J］. Aiaa Journal, 2012, 2 (4): 32.

［9］ Love A E H. A treatise on the mathematical theory of elasticity ［M］. Cambridge: Cambridge University Press, 2013.

［10］ Hosseini-Hashemi S, Fadaee M. On the free vibration of moderately thick spherical shell panel—a new exact closed-form procedure ［J］. Journal of Sound and Vibration, 2011, 330 (17): 4352-4367.

［11］ Wang X H, Redekop D. Natural frequencies analysis of moderately-thick and thick toroidal shells ［J］. Procedia Engineering, 2011, 14 (2259): 636-640.

［12］ Dong W, Liu Y, Xiang Z, et al. An analytical method for free vibration analysis of functionally graded beams with edge cracks ［J］. Journal of Sound & Vibration, 2012, 331 (7): 1686-1700.

［13］ Koiter W T. Buckling and post-buckling behavior of a cylindrical panel under axial compression ［R］. Amsterdam: Report S 476 National Aeronautical Research Institute, 1956.

［14］ Wei Y, Xiao K Y, Tong Y K, et al. Summary for ultimate bearing capacity research methods of engineering-structure containing flaws ［J］. Shanxi Architecture, 2014.

［15］ Aydin K. Free vibration of functionally graded beams with arbitrary number of surface cracks ［J］. European Journal of Mechanics, 2013, 42 (42): 112-124.

［16］ Yoon H I, Son I S, Ahn S J, et al. Free vibration analysis of Euler-Bernoulli beam with double cracks ［J］. Journal of Mechanical Science and Technology, 2007, 21: 476-485.

［17］ Jafari A A, Bagheri M. Free vibration of non-uniformly ring stiffened cylindrical shells using analytical, experimental and numerical methods ［J］. Thin-Walled Structures, 2006, 44 (1): 82-90.

［18］ Jaan Lellep, Larissa and Roots. Vibrations of cylindrical shells with circumferential cracks ［J］. Wseas Transactions on Mathematics, 2010, 9 (9): 689-699.

［19］ Jin C C, Zhu x, Li T, et al. Coupled vibration feature analysis for a fluid-filled cylindrical shell with a circumferential surface crack ［J］. Journal of Vibration and Shock, 2018, 37 (23): 71-77.

［20］ Lee H W, Kwak M K. Free vibration analysis of a circular cylindrical shell using the rayleigh-ritz method and comparison of different shell theories ［J］. Journal of Sound and Vibration, 2015, 353: 344-377.

［21］ Roots L. Non-axisymmetric Vibrations of Stepped Cylindrical Shells Containing Cracks ［J］. Seoul: International Conference on the Mechanics of Biological Systems and Materials, 2014.

［22］ Yin T, Lam Heung-Fai. Dynamic analysis of finite-length circular cylindrical shells with a circumferential surface crack ［J］. Journal of Engineering Mechanics, 2013, 139 (10): 1419-1434.

［23］ Dehghani O S, Esmaeilpour E H, Vafaei A H, et al. Free vibration of cracked cylindrical shells ［J］., 2008.

［24］ Zhu X, Li T Y, Zhao Y, et al. Vibrational power flow analysis of thin cylindrical shell with a circumferential surface crack ［J］. Journal of Sound & Vibration, 2007, 302 (1-2): 323-349.

［25］ Zheng S, Yu Y, Qiu M, et al. A modal analysis of vibration response of a cracked fluid-filled cylindrical shell ［J］. Applied Mathematical Modelling, 2021, 91: 934-958.

［26］ Chen X D, Ye K S. Analysis of free vibration of moderately thick circular cylindrical shells using the dynamic stiffness method ［J］. Journal of Vibration and Shock, 2016, 35 (6): 84-90.

［27］ Chen J S, Zhang Y. The analysis of the limit element method on the free vibration of the cylinder shell

[J]. Machinery Design & Manufacture, 2006, (11): 16-17.

[28] Flugge. Stresses in shells [J]. Springer., 1973.

[29] Wang Z Q, Huang L H, Li X B, et al. Vibration analysis of circular cylindrical shells under arbitrary boundary conditions [J]. Ship Science and Technology, 2017, 39 (4): 24-29.

[30] Li Q S. Free vibration analysis of non-uniform beams with an arbitrary number of cracks and concentrated masses [J]. Journal of Sound and Vibration, 2001, 252 (3): 509-525.

[31] Grigorenko Ya, A Efimova, T L, et al. On one approach to studying free vibrations of cylindrical shells of variable thickness in the circumferential direction within a refined statement [J]. Journal of Mathematical Sciences, 2010, 171 (4): 548-563.

[32] Grigorenko A Y, Puzyrev S V, Prigoda A P, et al. Theoretical-experimental investigation of frequencies of free vibrations of circular cylindrical shells [J]. Journal of Mathematical Sciences, 2011, 174 (2): 254-267.

[33] Pellicano F. Vibrations of circular cylindrical shells: theory and experiments [J]. Journal of Sound & Vibration, 2007, 303 (1-2): 154-170.

[34] Babuska I, Rheinboldt W C. Adaptive approaches and reliability estimates in finite element analysis [J]. Computer Methods in Applied Mechanics and Engineering, 1979, 17 (3): 519-540.

[35] Zienkiewicz O C, Zhu J Z. The superconvergent patch recovery (SPR) and adaptive finite element refinement [J]. Computer Methods in Applied Mechanics and Engineering, 1992, 101 (1-3): 207-224.

[36] Zienkiewicz O C. The background of error estimation and adaptivity in finite element computations [J]. Computer Methods in Applied Mechanics and Engineering, 2006, 195 (4-6): 207-213.

[37] Bespalov A, Haberl A, Praetorius D, et al. Adaptive FEM with coarse initial mesh guarantees optimal convergence rates for compactly perturbed elliptic problems [J]. Computer Methods in Applied Mechanics and Engineering, 2017, 317: 318-340.

[38] Oden J T, Babuska I, Baumann C E, et al. A discontinuous hp finite element method for diffusion problems [J]. Journal of Computational Physics, 1998, 146 (2): 491-519.

[39] Wang Y. Adaptive analysis of damage and fracture in rock with multiphysical fields coupling, [M]. Springer Press, 2021.

[40] Arthurs C J, Bishop M J, Kay D, et al. Efficient simulation of cardiac electrical propagation using high-order finite elements II: adaptive p-version [J]. Journal of Computational Physics, 2013, 253: 443-470.

[41] Wiberg N E, Hager P. Error estimation and adaptivity for h-version eigenfrequency analysis [J]. Studies in Applied Mechanics, 1998, 47: 461-475.

[42] Wang Y L. An h-version adaptive FEM for eigenproblems in system of second order ODEs: vector Sturm-Liouville problems and free vibration of curved beams [J]. Engineering Computations, 2020, 37 (1): 1210-1225.

[43] Gomez-Revuelto I, Garcia-Castillo L E, Llorente-Romano S, et al. A three-dimensional self-adaptive hp finite element method for the characterization of waveguide discontinuities [J]. Computer Methods in Applied Mechanics and Engineering, 2012, 249: 62-74.

[44] Oden J T, Duarte C, Zienkiewicz O C, et al. A new cloud-based hp finite element method [J]. Computer Methods in Applied Mechanics & Engineering, 1998, 153 (1-2): 117-126.

[45] Wang Y L, Wang J H. An hp-version adaptive finite element algorithm for eigensolutions of moderately thick circular cylindrical shells via error homogenisation and higher-order interpolation [J].

Engineering Computations, 2021.

[46] Wang Y L. Adaptive finite element-discrete element analysis for striatal movement and microseismic behaviors induced by multistage propagation of three-dimensional multiple hydraulic fractures [J]. Engineering Computations, 2020, 38 (5): 1350-1371.

[47] Wang Y L, Ju Y, Yang Y M, et al. Adaptive finite element-discrete element analysis for microseismic modelling of hydraulic fracture propagation of perforation in horizontal well considering pre-existing fractures [J]. Shock and Vibration, 2018, : 1-14.

[48] Wang Y L, Ju Y, Chen J L, et al. Adaptive finite element-discrete element analysis for the multistage supercritical CO2 fracturing of horizontal wells in tight reservoirs considering pre-existing fractures and thermal-hydro-mechanical coupling [J]. Journal of Natural Gas Science and Engineering, 2019, 61: 251-269.

[49] Babuska I, Rheinboldt W C. A-posteriori error estimates for the finite element method [J]. International Journal for Numerical Methods in Engineering, 1978, 12: 1597-1615.

[50] Ainsworth M, Oden J T. A procedure for a posteriori error estimation for h-p finite element methods [J]. Computer Methods in Applied Mechanics and Engineering, 1992, 101: 73-96.

[51] Ainsworth M, Oden J T. A unified approach to a posteriori error estimation using element residual methods [J]. Numerische Mathematik, 1993, 65: 23-50.

[52] Yin T, Li Q S, Zhu H P, et al. A new solution method for vibration analysis of circular cylindrical thin shells with a circumferential surface crack. [J]. Advanced Materials Research, 2013, 639-640 (1): 1003-1009.

[53] Arndt M, Machado R D, Scremin A, et al. An adaptive generalized finite element method applied to free vibration analysis of straight bars and trusses [J]. Journal of Sound and Vibration, 2010, 329 (6): 659-672.

[54] Schillinger D, Rank E. An unfitted hp-adaptive finite element method based on hierarchical B-splines for interface problems of complex geometry [J]. Computer Methods in Applied Mechanics and Engineering, 2011, 200 (47-48): 3358-3380.

[55] Bao G, Hu G, Liu D, et al. An h-adaptive finite element solver for the calculations of the electronic structures [J]. Journal of Computational Physics, 2012, 231 (14): 4967-4979.

[56] Wiberg N E, Bausys R, Hager P, et al. Improved eigenfrequencies and eigenmodes in free vibration analysis [J]. Computers and Structures, 1999, 73 (1-5): 79-89.

[57] Wiberg N E, Bausys R, Hager P, et al. Adaptive h-version eigenfrequency analysis [J]. Computers and Structures, 1999, 71 (5): 565-584.

[58] Wang Y L, Ju Y, Zhuang Z, et al. Adaptive finite element analysis for damage detection of non-uniform Euler-Bernoulli beams with multiple cracks based on natural frequencies [J]. Engineering Computations, 2018, 35 (3): 1203-1229.

[59] Greenberg L. A prüfer method for calculating eigenvalues of self-adjoint systems of ordinary differential equations: parts 1 and 2 [R]. Technical Report TR91-24, 1991, University of Maryland at College Park MD.

[60] Kurochkin S V. Indexing of eigenvalues of boundary value problems for Hamiltonian systems of ordinary differential equations [J]. Computational Mathematics and Mathematical Physics, 2014, 54 (3): 439-442.

[61] Zienkiewicz O C, Taylor R L, J, et al. The finite element method: its basis and fundamentals [M]. 7th Edition. Oxford: Elsevier (Singapore) Pte Ltd., 2015.

［62］ Wilkinson J H. The Algebraic Eigenvalue Problem ［M］. Oxford: Clarendon Press, 1965.

［63］ Wilkinson J H, Reinsch C. Linear algebra, handbook for automatic computation ［M］. New York: Springer-Verlag, 1971.

Chapter 10
Summary and prospect

10. 1　Summary

The chapters of the book can be summarized as follows:

(1) In Chapter 1, the research background and significances of eigenproblems in vibration, stability, and damage disturbance of beam and shell containing and high-performance adaptive finite element method are well summarized and analysed. This section can provide the research aims and contents of the book.

(2) In Chapter 2, an h-version adaptive finite element method based on the superconvergent patch recovery displacement method for eigenproblems in system of second order ordinary differential equations (ODEs) was presented. The high-order shape function interpolation technique is further introduced to acquire superconvergent solution of eigenfunction, and superconvergent solution of eigenvalue is obtained by computing the Rayleigh quotient. Superconvergent solution of eigenfunction is used to estimate the error of finite element solution in energy norm. The mesh is then subdivided to generate an improved mesh, based on the error. Representative eigenproblems examples, containing typical vector Sturm-Liouville (SL) and free vibration of beams problems involved aforementioned challenging issues, are selected to evaluate the accuracy and reliability of the proposed method. Non-uniform refined meshes are established to suit eigenfunctions change, and numerical solutions satisfy the pre-specified error tolerance.

(3) In Chapter 3, an adaptive method was proposed to analyse the in-plane and out-of-plane free vibrations of the variable geometrical Timoshenko beams. In the post-processing stage of the displacement-based finite element method, the superconvergent patch recovery method and high-order shape function interpolation technique were used to obtain the superconvergent solution of mode (displacement). The superconvergent solution of mode was used to estimate the error of the finite element solution of mode in the energy form under the current mesh. Furthermore, an adaptive mesh refinement was proposed by mesh subdivision to derive an optimised mesh and accurate finite element solution to meet the preset error tolerance. The results computed using the proposed algorithm were in good agreement with those computed using other high-precision algorithms, thus validating the accuracy of the proposed algorithm for beam analysis. The numerical analysis of parabolic curved beams, beams with variable cross-sections and curvatures, elliptically curved beams, and circularly curved beams helped verify that the solutions of frequencies were consistent with the results obtained using other specially developed methods. The proposed method is well suited for the mesh adaptive analysis of a curved beam structure for analysing the changes in high-order vibration mode. The parts where the vibration mode changed significantly were locally densified; a relatively fine mesh division was adopted, which validated the reliability of the mesh optimisation processing of the proposed algorithm.

(4) In Chapter 4, a scheme for cross-sectional damage defects in a circularly curved beam was established to simulate the depth, location, and the number of multiple cracks by implementing cross-section reduction induced by microcrack damage. In addition, the h-version finite element mesh adaptive analysis method of the variable cross-section Timoshenko beam was developed. The superconvergent solution of the vibration mode of the cracked curved beam was obtained using the superconvergent patch recovery displacement method to determine the finite element solution. The superconvergent solution of the frequency was obtained by computing the Rayleigh quotient. The superconvergent solution of the eigenfunction was used to estimate the error of the finite element solution in the energy norm. The mesh was then subdivided to generate an improved mesh based on the error. Accordingly, the final optimised meshes and high-precision solution of natural frequency and mode shape satisfying the preset error tolerance can be obtained. Lastly, the disturbance behaviour of multi-crack damage on the vibration mode of a circularly curved beam was also studied. Numerical results of the free vibration and damage disturbance of cracked curved beams with cracks were obtained. The influences of crack damage depth, crack damage number, and crack damage distribution on the natural frequency and mode of vibration of a circularly curved beam were quantitatively analysed. Numerical examples indicate that the vibration mode and frequency of the beam would be disturbed in the region close to the crack damage, and a greater crack depth translates to a larger frequency change. For multi-crack beams, the number and distribution of cracks also affect the vibration mode and natural frequency. The adaptive method can use a relatively dense mesh near the crack to adapt to the change in the vibration mode near the crack, thus verifying the efficacy, accuracy, and reliability of the method.

(5) In Chapter 5, the weakened properties analogy was used to describe cracks in this model. The adaptive strategy proposed in this paper provides accurate, efficient, and reliable eigensolutions of frequency and mode (i. e. eigenpairs as eigenvalue and eigenfunction) for Euler-Bernoulli beams with multiple cracks. Based on the frequency measurement method for damage detection, utilizing the difference between the actual and computed frequencies of cracked beams, the inverse eigenproblems are solved iteratively for identifying the residuals of locations and sizes of the cracks by the Newton-Raphson iteration technique. In the crack detection, the estimated residuals are added to obtain reliable results, which is an iteration process that will be expedited by more accurate frequency solutions based on the proposed method for free vibration problems. Numerical results are presented for free vibration problems and damage detection problems of representative non-uniform and geometrically stepped Euler-Bernoulli beams with multiple cracks to demonstrate the effectiveness, efficiency, accuracy and reliability of the proposed method.

(6) In Chapter 6, for the elastic buckling of circular curved beams with cracks, the

section damage defect analogy scheme of a circular arc curved beam crack was established to simulate the crack size (depth), position, and number. The h-version finite element mesh adaptive analysis method of the variable section Euler-Bernoulli beam was introduced to solve the elastic buckling problem of circular arc curved beams with crack damage. The optimised mesh and high-precision buckling load and buckling mode solutions satisfying the preset error limit were obtained. The results of testing typical examples show that (1) the established section damage defect analogy scheme of circular arc curved beam crack can effectively realise the simulation of crack size (depth), position, and number. The solution strictly satisfies the preset error limit; (2) the non-uniform mesh refinement in the algorithm can be adapted to solve the arbitrary order frequencies and modes of cracked cylindrical shells under the conditions of different ring wave numbers, crack positions, and crack depths; and (3) the change in the buckling mode caused by crack damage is applicable to the study of elastic buckling under various curved beam angles and crack damage distribution conditions. The influence of the degree of crack damage on the buckling load and buckling mode of a circular arc curved beam was quantitatively analysed, and the accuracy and reliability of the algorithm were tested.

(7) In Chapter 7, on a given finite element mesh, the solutions of the frequency mode of the moderately thick circular cylindrical shell were obtained using the conventional finite element method. Subsequently, the superconvergent patch recovery displacement method and high-order shape function interpolation techniques were introduced to obtain the superconvergent solution of the mode (displacement), while the superconvergent solution of the frequency was obtained using the Rayleigh quotient computation. Finally, the superconvergent solution of the mode was used to estimate the errors of the finite element solutions in the energy norm, and the mesh was subdivided to generate a new mesh in accordance with the errors. In this study, a high-precision and reliable superconvergent patch recovery solution for the vibration modes of variable geometrical rotating cylindrical shells was developed. Different circumferential wave numbers, and thickness-length ratios, the optimised finite element meshes, and high-precision solutions satisfying the preset error limits were obtained successfully to solve the frequency and mode of continuous orders of rotating cylindrical shells with multiple boundary conditions such as simple and fixed supports, demonstrating good solution efficiency. The existing problem on the difficulty of adapting a set of meshes to the changes in vibration modes of different orders is finally overcome by applying the adaptive optimisation.

(8) In Chapter 8, an hp-version adaptive finite element algorithm was proposed for determining the eigensolutions of the free vibration of moderately thick circular cylindrical shells via error homogenisation and higher-order interpolation. This algorithm first develops the established h-version mesh refinement method for detecting the non-uniform distributed optimised meshes, where the error estimation and element subdivision approaches based on the superconvergent patch recovery displacement method are

introduced to obtain high-precision solutions. The errors in the vibration mode solutions in the global space domain are homogenised and approximately the same. Subsequently, on the refined meshes, the algorithm uses higher-order shape functions for the interpolation of trial displacement functions to reduce the errors quickly, until the solution meets a pre-specified error tolerance condition. In this algorithm, the non-uniform mesh generation and higher-order interpolation of shape functions are suitable for addressing the problem of complex frequencies and modes caused by variable structural geometries. Numerical results are presented for moderately thick circular cylindrical shells with different geometrical factors (circumferential wave number, thickness-to-radius ratio, thickness-to-length ratio) to demonstrate the effectiveness, accuracy, and reliability of the proposed method. The hp-version refinement uses fewer optimised meshes than h-version mesh refinement, and only one-step interpolation of the higher-order shape function yields the eigensolutions satisfying the accuracy requirement.

(9) In Chapter 9, an adaptive finite element method and a crack damage characterisation method for moderately thick circular cylindrical shells were developed. By introducing the inverse power iteration method, error estimation, and mesh subdivision refinement technique for the analysis of finite element eigenvalue problems, an adaptive computation scheme was constructed for the free vibration problem of moderately thick circular cylindrical shells with circumferential crack damage. Based on typical numerical examples, the established adaptive finite element solution for the free vibration of moderately thick circular cylindrical shells demonstrated its suitability for solving the high-precision free vibration frequency and mode of cylindrical shell structures. The any order frequency and mode shape of cracked cylindrical shells under the conditions of different ring wave numbers, crack locations, crack depths, and multiple cracks were successfully solved. The influences of the location, depth, and number of cracks on the disturbance of dynamic behaviours were analysed.

(10) In Chapter 10, all chapters in the book are summarized, and the prospect for future work is introduced.

10. 2　Prospect

Based on the research work presented in this book, further work can be carried out:

(1) **Adaptive algorithm theory for three-dimensional dynamic analysis based on superconvergent patch recovery method.** In this study, the superconvergent point is used to obtain the high-precision solutions in global domain, and the high-precision solutions are used to estimate the error of the solutions, so as to realize the adaptive solution process. The next research focus is to realize the high-precision solution based on the superconvergent patch recovery in three-dimensional structure, and use the solution to establish an adaptive algorithm. For three-dimensional dynamic problems, if the field-scale

model is set to be three-dimensional, more superconvergent points will appear, and more efficient patch recovery algorithm (such as deep learning) need to be developed.

（2）**Combined adaptive mesh refinement and continuum-discontinuum analysis for structures and rocks.** To characterize the evolution of continuous stress fields and discrete fracture fields in structures and deep reservoir rocks, it is necessary to establish numerical methods of continuum-discontinuum analysis and develop adaptive finite element algorithms. The high-precision solutions of this algorithm should satisfy a pre-set tolerance and ensure an efficient computational process via mesh refinement. To characterize the three-dimensional occurrence state and the non-planar damage evolution and crack propagation behaviours of deep reservoir rock, a three-dimensional model to simulate structure and rock behaviours is needed and the need to establish a three-dimensional adaptive algorithm has become urgent.

（3）**High-performance parallel computing software for large-scale structural engineering problems.** To determine the solutions of three-dimensional structural eigenproblems, using the three-dimensional adaptive algorithm, a large amount of computation is needed. Thus, higher efficiency in computing is required. An important way to achieve this is to introduce parallel computing methods to enable multicore and multithreaded parallel computing. To develop this proposed program into a computing platform that can be used by all kinds of users, it is necessary to develop a user-friendly computing software that is suitable for the computations and analyses comprising large-scale structural engineering problems.